Fractal Analysis of Breast Masses in Mammograms

Synthesis Lectures on Biomedical Engineering

Editor
John D. Enderle, *University of Connecticut*

Lectures in Biomedical Engineering will be comprised of 75- to 150-page publications on advanced and state-of-the-art topics that span the field of biomedical engineering, from the atom and molecule to large diagnostic equipment. Each lecture covers, for that topic, the fundamental principles in a unified manner, develops underlying concepts needed for sequential material, and progresses to more advanced topics. Computer software and multimedia, when appropriate and available, are included for simulation, computation, visualization and design. The authors selected to write the lectures are leading experts on the subject who have extensive background in theory, application and design.

The series is designed to meet the demands of the 21st century technology and the rapid advancements in the all-encompassing field of biomedical engineering that includes biochemical, biomaterials, biomechanics, bioinstrumentation, physiological modeling, biosignal processing, bioinformatics, biocomplexity, medical and molecular imaging, rehabilitation engineering, biomimetic nano-electrokinetics, biosensors, biotechnology, clinical engineering, biomedical devices, drug discovery and delivery systems, tissue engineering, proteomics, functional genomics, and molecular and cellular engineering.

Fractal Analysis of Breast Masses in Mammograms
Thanh M. Cabral and Rangaraj M. Rangayyan
2012

Medical Equipment Maintenance: Management and Oversight
Binseng Wang
2012

Capstone Design Courses, Part II: Preparing Biomedical Engineers for the Real World
Jay R. Goldberg
2012

Chronobioengineering: Introduction to Biological Rhythms with Applications
Donald McEachron
2012

Phonocardiography Signal Processing
Abbas K. Abbas and Rasha Bassam
2009

Introduction to Biomedical Engineering: Biomechanics and Bioelectricity - Part II
Douglas A. Christensen
2009

Introduction to Biomedical Engineering: Biomechanics and Bioelectricity - Part I
Douglas A. Christensen
2009

Landmarking and Segmentation of 3D CT Images
Shantanu Banik, Rangaraj M. Rangayyan, and Graham S. Boag
2009

Basic Feedback Controls in Biomedicine
Charles S. Lessard
2009

Understanding Atrial Fibrillation: The Signal Processing Contribution, Part I
Luca Mainardi, Leif Sörnmo, and Sergio Cerutti
2008

Understanding Atrial Fibrillation: The Signal Processing Contribution, Part II
Luca Mainardi, Leif Sörnmo, and Sergio Cerutti
2008

Introductory Medical Imaging
A. A. Bharath
2008

Lung Sounds: An Advanced Signal Processing Perspective
Leontios J. Hadjileontiadis
2008

An Outline of Informational Genetics
Gérard Battail
2008

Neural Interfacing: Forging the Human-Machine Connection
Susanne D. Coates
2008

Fractal Analysis of Breast Masses in Mammograms

Thanh M. Cabral and Rangaraj M. Rangayyan

ISBN-13: 978-3-031-00526-8   paperback
ISBN-13: 978-3-031-01654-7   ebook

DOI 10.1007/978-3-031-01654-7

A Publication in the Springer series
SYNTHESIS LECTURES ON BIOMEDICAL ENGINEERING

Lecture #46
Series Editor: John D. Enderle, *University of Connecticut*
Series ISSN
Synthesis Lectures on Biomedical Engineering
Print 1930-0328   Electronic 1930-0336

Fractal Analysis of Breast Masses in Mammograms

Thanh M. Cabral and Rangaraj M. Rangayyan
University of Calgary, Calgary, Alberta, Canada

SYNTHESIS LECTURES ON BIOMEDICAL ENGINEERING #46

ABSTRACT

Fractal analysis is useful in digital image processing for the characterization of shape roughness and gray-scale texture or complexity. Breast masses present shape and gray-scale characteristics in mammograms that vary between benign masses and malignant tumors. This book demonstrates the use of fractal analysis to classify breast masses as benign masses or malignant tumors based on the irregularity exhibited in their contours and the gray-scale variability exhibited in their mammographic images. A few different approaches are described to estimate the fractal dimension (FD) of the contour of a mass, including the ruler method, box-counting method, and the power spectral analysis (PSA) method. Procedures are also described for the estimation of the FD of the gray-scale image of a mass using the blanket method and the PSA method.

To facilitate comparative analysis of FD as a feature for pattern classification of breast masses, several other shape features and texture measures are described in the book. The shape features described include compactness, spiculation index, fractional concavity, and Fourier factor. The texture measures described are statistical measures derived from the gray-level cooccurrence matrix of the given image. Texture measures reveal properties about the spatial distribution of the gray levels in the given image; therefore, the performance of texture measures may be dependent on the resolution of the image. For this reason, an analysis of the effect of spatial resolution or pixel size on texture measures in the classification of breast masses is presented in the book.

The results demonstrated in the book indicate that fractal analysis is more suitable for characterization of the shape than the gray-level variations of breast masses, with area under the receiver operating characteristics of up to 0.93 with a dataset of 111 mammographic images of masses. The methods and results presented in the book are useful for computer-aided diagnosis of breast cancer.

KEYWORDS

blanket method, box-counting method, breast cancer, breast masses, compactness, computer-aided diagnosis, digital image processing, Fourier descriptors, fractal analysis, fractal dimension, fractional concavity, mammography, pattern classification, pattern recognition, power spectral analysis, ruler method, shape analysis, signature of a contour, spiculation index, texture analysis

Dedicated to
Ba, Mẹ, and Kevin

Thanh

Contents

Preface

Whereas linear filters and models have played well-established roles in the analysis of signals and images, there has been increasing interest in nonlinear and nonstationary methods to analyze, characterize, and understand increasingly complex systems and signals. The substantial variations that are encountered in medical images within a given category as well as the overlap that exists in their characteristics across multiple categories call for sophisticated approaches for their analysis and classification. Fractal analysis is a relatively newly developed methodology that could assist in such situations.

Fractals have been observed to occur widely in nature. Several biological systems exhibit fractal behavior and patterns. However, there exist several different notions and models within the realm of fractals. The objective of this book is to explore the use of fractal analysis in discriminating between breast masses due to benign disease and malignant tumors caused by breast cancer, as seen in mammographic images. Although the methods described are based on profound mathematical models and theories, they may be understood by an attentive reader with a good training in mathematics at the level of first-year university. The large number of illustrations provided in the book should assist in gaining a good understanding of the concepts and methods involved as well as in appreciating their application to a real-life problem.

Notwithstanding the specific medical application considered in the present work, it is expected that the methods described should be useful in several other applications of imaging and image analysis.

We wish the reader a challenging exploration of the concepts and applications of fractals.

Thanh M. Cabral
Rangaraj M. Rangayyan
October 2012

Acknowledgments

We thank our collaborators who have contributed to parts of the research work leading to the present book: Dr. J. E. Leo Desautels, Dr. Liang Shen, Dr. Hilary Alto, Dr. Fábio J. Ayres, Shormistha Prajna, Dr. Shantanu Banik, Faraz Oloumi, and Dr. Mauro Tambasco at the University of Calgary, and Professor Asoke K. Nandi, The University of Liverpool, Liverpool, U.K.

We thank the Natural Sciences and Engineering Research Council of Canada, Alberta Heritage Foundation for Medical Research, and Canadian Breast Cancer Foundation: Prairies and Northwest Territories Chapter, for supporting our projects.

Thanh M. Cabral
Rangaraj M. Rangayyan
October 2012

List of Symbols and Abbreviations

ANOVA	analysis of variance
AUC	area under the ROC curve
bpp	bits per pixel
BSE	breast self-examination
C	compactness
CAD	computer-aided diagnosis
CB	circumscribed benign
CM	circumscribed malignant
fBm	fractional Brownian motion
F_{cc}	fractional concavity
FD	fractal dimension
FF	Fourier factor
FLDA	Fisher linear discriminant analysis
FN	false negative
FNR	false-negative rate
FP	false positive
FPR	false-positive rate
GLCM	gray-level cooccurrence matrix
H	Hurst exponent
LOO	leave-one-out
p	p-value
PPV	positive-predictive value
PSA	power spectral analysis
PSD	power spectral density
RBST	rubber band straightening transform
ROC	receiver operating characteristics
ROI	region of interest
SB	spiculated benign
SI	spiculation index
SM	spiculated malignant
T_c	T-critical

t_s	t-statistic
TN	true negative
TNR	true-negative rate
TP	true positive
TPR	true-positive rate
1D	one-dimensional
2D	two-dimensional
3D	three-dimensional

CHAPTER 1

Computer-Aided Diagnosis of Breast Cancer

1.1 MAMMOGRAPHY

Early detection and prompt treatment are key steps toward reducing the mortality rate due to breast cancer. The most common symptom of breast cancer is the presence of a lump or a mass in the breast tissue. In the past, breast self-examination (BSE) was promoted as an effective method for detecting early symptoms of breast cancer. However, recent studies have shown inadequate evidence that routine, systematic BSE reduces deaths due to breast cancer [1]. With the advancement in electronics and technology over the past 50 years, numerous new imaging modalities have been developed as noninvasive means to visualize and assess the human body for diseases [2].

The most effective imaging modality for the breast is mammography [3]. A mammogram is an X-ray image of the breast. The various levels of gray on a mammogram are associated with the net density of tissues along the paths of the X rays through the breast. The denser the tissue, the more X-ray photons are absorbed or attenuated and the lighter the corresponding region appears on the mammogram. Signs of abnormalities in the breast can usually be seen in a mammogram well before the abnormalities can be felt in a physical examination of the breast. Mammography is currently the best screening method available for the detection of breast cancer at its earliest stages when the disease often exhibits no palpable symptoms. It is the only method that has been proven in multiple randomized clinical studies to decrease mortality due to breast cancer. The Canadian Cancer Society recommends women from 50 to 69 years of age to participate in regular mammographic screening, once every two years [4].

Evidence of abnormality in the breast is usually indicated by the presence of microcalcifications, dense regions or masses, bilateral asymmetry between the left and right breasts, and architectural distortion. Breast microcalcifications are deposits of calcium that appear as white specks on mammograms. Calcium has a higher X-ray attenuation coefficient as compared to soft tissue; thus, calcium appears as a bright region on a mammogram. Macrocalcifications (larger than 0.5 mm in diameter) are generally benign. Microcalcifications are classified as more likely to be malignant or more likely to be benign based on the arrangement of clusters of microcalcifications, the amount or number of microcalcifications in a specified area, and the changes in the pattern of clusters of microcalcifications compared to past mammograms of the same breast. In general, irregularly shaped calcifications that do not orient toward the nipple are likely to be associated with malignancy, whereas round and oval calcifications that are generally uniform in size are likely to be associated with benign breast disease [3].

A dense region or mass, depending on its morphological features, may represent a localized sign of breast cancer. A suspicious region in a mammogram is generally categorized as either benign or malignant. A typical benign mass has a well-defined boundary and is round or oval in shape, whereas a malignant tumor usually has a fuzzy boundary and is irregular in shape [3].

Architectural distortion is the distortion of the normal architecture of the breast parenchyma with no definite mass visible. Architectural distortion may be a subtle indication of an early stage of development of malignancy.

Bilateral asymmetry is a condition of asymmetrical or substantially different organization of breast tissue patterns between images of the left and right breasts of an individual. The appearance of asymmetry may be indicative of a developing mass, pathological displacement of tissue, or merely poor technique during image acquisition.

The presence of an abnormal condition in the breast indicated by the mammographic patterns described above could be a sign of cancer, precancerous cell formation, benign breast disease, trauma, or other conditions. The intent during screening or diagnosis is to identify and classify such abnormal regions.

Limitations exist in screening mammography. Bird et al. [5] suggested that, occasionally, signs of abnormalities on mammograms go undetected by the radiologists because they were either overlooked or misinterpreted. Overlooking errors accounted for 43% of all errors, whereas misinterpretation errors accounted for 52% of all errors in the study by Bird et al. [5]. In addition, interpretations of screening mammograms among radiologists vary due to differing levels of expertise. Variability also exists within the performance of a radiologist as a result of distraction and fatigue that may be caused by the high volume of work, or inconsistent application of knowledge. Regardless, a fundamental limitation of mammography is caused by the fact that the image obtained is a two-dimensional (2D) projection of the three-dimensional (3D) breast in which tissues at different depths are superimposed.

The benefit of double reading over single reading of screening mammograms has been extensively investigated. Double reading implies that a mammogram is interpreted by two radiologists and the results are combined in some manner to arrive at the final decision. Increasing the number of radiologists to read screening mammograms independently has been shown to increase the detection rate of breast abnormalities without any significant increase in the recall rate [6]. Harvey et al. [7] reported that double reading increased the number of cancers found by 6.3% and caused an additional 1.5% of the total number of women in the study to be recalled. Ciatto et al. [8] suggested that double reading of screening mammograms improved the detection of difficult cancers and large tumors that were missed due to fatigue and distraction during the first reading. A review by Elmore and Brenner [9] suggested that double reading by individuals with complementary visual search patterns may offer the best outcome between the detection rate and the recall rate, as compared to double reading by individuals with similar visual search patterns. In general, double reading of screening mammograms improves the rate of detection of cancer by 5% to 15% as compared to single reading.

In modernized screening and diagnostic centers, digital mammography is used instead of traditional screen-film mammography. Screen-film mammography and digital mammography both use X rays to produce an image of the breast; the difference is that traditional mammography uses a screen-film receptor to capture the information and the image is produced after the film is processed, whereas in digital mammography, a solid-state detector converts the X rays to digital data and immediately produces the image on a computer monitor. Digital mammograms are easy to transmit, retrieve, and store in a database. The major benefits of digital mammography are that the image can be manipulated electronically to magnify a region of interest (ROI), change the contrast, and alter the brightness. Recently, a study by Pisano et al. [10] found digital mammography to be more accurate than film mammography for imaging dense breasts, the usual situation in women younger than 50 years of age. More importantly, a digital mammogram or a digitized film mammogram can be submitted to a computer system to be processed, analyzed, and interpreted; such an approach is known as computer-aided detection or diagnosis (CADe, CADx, or CAD).

1.2   COMPUTER-AIDED DIAGNOSIS OF BREAST CANCER

Heightened awareness of breast cancer has led large populations of asymptomatic women to participate in regular mammographic screening programs [11]. While the demand for mammographic services is rising, the interest among radiologists in interpreting mammograms is decreasing, according to Bassett et al. [12]. The approach of double reading screening mammograms becomes inefficient and impractical in some regions. With the large volumes of mammograms being acquired and the availability of limited numbers of radiologists to interpret them arises the need to enlist help from computers to analyze mammograms more accurately and efficiently. Early studies in the 1960s on automated computer-based diagnosis in radiology concluded that computers could not replace the expertise of experienced radiologists [13]. In the 1980s, the focus shifted to employing computers as tools to assist radiologists in their decision making process. Hence, CAD systems became a major research topic in diagnostic imaging [13].

A CAD system is built upon complex computer algorithms. The general framework of a CAD scheme for breast cancer involves image acquisition, preprocessing, segmentation, analysis, and classification. Mammographic images may be acquired using conventional screen-film technology or newer digital technology. Digital mammography has advantages over screen-film mammography in terms of storage, transmission, retrieval, and manipulation of the images. A CAD system requires digital data in the form of digitized film mammograms or digital mammograms. Mammograms are initially preprocessed to enhance desirable features while reducing noise and artifacts. After preprocessing, the mammograms are partitioned into various segments. Segmentation in mammography involves separating the breast region from the background tissues (such as the pectoral muscle), identifying the regions suspected to contain abnormalities, and obtaining the borders of suspicious regions, if possible. Suspicious regions are characterized with features that help in discriminating between normal tissues and diseased or cancerous tissues. Feature selection is applied to select the most relevant features while discarding irrelevant and redundant features to create an optimal subset of features. Finally, methods for pattern classification are applied to categorize the regions into cancerous, diseased, or normal classes. CAD systems are designed to assist radiologists by prompting or alerting them to suspicious regions on mammograms that may have been missed during their first reading.

The quality of screening mammography is analyzed using measures such as sensitivity, specificity, false-positive rate (FPR), recall rate, and the positive-predictive value (PPV). Sensitivity is the proportion of breast cancer cases correctly identified out of the total number of breast cancer cases provided. Specificity is the proportion of the normal cases correctly identified as normal out of the total number of normal cases provided. FPR is the proportion of the normal cases incorrectly identified as cancer or abnormal out of the total number of normal cases. The recall rate is the rate at which screened women are asked to return for additional assessments. PPV is the proportion of the cases that are subsequently diagnosed to have cancer to those that were initially labeled as suspicious.

Introducing CAD technology to assist radiologists in interpreting screening mammograms has been shown to improve the radiologists' accuracy in the detection of cancer [14]. In the typical or recommended manner of operation, a radiologist working with the assistance of a CAD system performs the first reading of the mammogram and records the results. Then, the CAD system processes the image for suspicious regions, highlights such regions, and prompts for further review by the radiologist. In such an application, the CAD system functions as a second reader to alert the radiologist to regions that may have been overlooked. The final assessment of the mammogram, whether or not to recall the patient for further evaluation, is the decision of the radiologist. Two experienced breast radiologists, Freer and Ulissey [14], conducted a study on 12,860 screening mammograms over a 12-month period, to compare their combined performance in detecting breast cancer without and with CAD. The initial reading of each mammogram was performed by either Freer or Ulissey and the result was recorded. Next, the radiologist used the prompts from a CAD system and reevaluated the mammogram, and the result was recorded. Freer and Ulissey found that their interpretations with CAD resulted in a 19.5% increase in the number of detected cases of cancer. The PPV was the same by both methods, without and with CAD. The recall rate was higher with CAD (7.7%) compared to that without CAD (6.5%). The study also observed an increase from 73% to 78% in the proportion of detected cases of early-stage breast cancer. The radiologists concluded that CAD technology is valuable in screening mammography [14].

Other studies have suggested that single reading with CAD could be an alternative to double reading. A study by Gromet [15] indicated that single reading with CAD and double reading yielded similar sensitivity (90.4% vs. 88.0%) and PPV (3.9% vs. 3.7%). The recall rate was significantly lower for single reading with CAD (10.6%) compared to double reading (11.9%). Both single reading with CAD and double reading resulted in increased detection sensitivity with only a small increase in the recall rate compared to single reading. Gilbert et al. [16] found that single reading with CAD and double reading were comparable in sensitivity (87.2% vs. 87.7%) and PPV (18.0% vs. 21.1%).

However, the recall rate for single reading with CAD (3.9%) was slightly higher, but with statistical significance, compared to double reading (3.4%).

A highly debated limitation of CAD for breast cancer is the large number of false positives (FPs) that CAD systems produce. FP diagnoses create unnecessary recalls and biopsies in healthy patients, resulting in emotional and financial strain. Research has produced inconsistent findings on CAD's sensitivity and recall rate. Perhaps the most damaging blow to the reputation of CAD for breast cancer is a recent study by Fenton et al. [17]. This study, involving more than 429,000 mammograms, is considered to be the most comprehensive analysis of CAD in breast screening to date. Fenton et al. found that interpretation with CAD resulted in lower specificity and PPV, higher rate of biopsy, and insignificant increase in sensitivity. Based on the results, the study concluded that CAD reduced the overall accuracy of interpretation of screening mammograms and increased the rate of biopsy. However, it could be argued that the positive effect of finding more cancers outweighs the negative effect of increased recalls.

The inconsistent findings on the performance of CAD systems for breast cancer should not be discouraging, but rather encouraging, to researchers because they present opportunities for further research and improvements in this field.

1.3   OBJECTIVES AND ORGANIZATION OF THE BOOK

The primary objective of this book is to present fractal analysis as an approach for the classification of breast masses, as seen in digitized mammograms, using their contours and gray-scale texture [18, 19]. Even though fractal analysis has been widely used in the analysis of biomedical images, only a few studies have applied the method specifically to study and classify mammographic masses (as discussed in Section 4.1.4).

Fractals are irregular geometric patterns made up of sets of infinitely smaller but identical patterns [20]. Fractal theory gives insights into tumor morphology and provides a mathematical platform for the analysis of complex and irregular tumor patterns. The fractal dimension (FD) may be used as a quantitative measure of the complexity of the contour or boundary and gray-scale variability exhibited by an object. Thus, the aim of our study is to determine whether patterns of breast masses exhibit fractal geometry and whether the FD of breast masses could be used to classify them as benign or malignant.

The secondary focus of this book is to compare the classification performance of FD against previously developed shape and texture measures [21, 22, 23, 24, 25, 26, 27], as well as to compare the classification performance of various combinations of these measures.

Finally, this book also presents the results of an investigation of the effect of spatial resolution or pixel size on texture features derived from gray-level cooccurrence matrices (GLCMs) in the classification of mammographic breast lesions as benign masses or malignant tumors [19].

The contents of the book are organized into seven chapters, as follows.

Chapter 2 presents a literature review of methods for the detection and analysis of breast masses. Selected measures of shape and texture reported in previous studies and those used in the present work are discussed in detail in this chapter.

Chapter 3 provides information about the datasets of images of breast masses used in this study.

Chapter 4 gives an explanation of the theory of fractals and the application of fractal analysis to characterize or quantify tumor patterns. This chapter presents methods to compute the FD of a breast lesion based on the shape of the breast lesion and the gray-level variation within the lesion.

Chapter 5 presents a brief discussion on the methods of pattern classification and analysis used in the present study. The methods include Fisher linear discriminant analysis (FLDA), the Bayesian classifier, and receiver operating characteristics (ROC).

Chapter 6 presents the results of classification using shape and texture features of breast masses. This chapter also presents the results of the investigation of the effect of pixel size on GLCM-based texture features in the classification of breast masses.

Finally, Chapter 7 summarizes the findings presented in this book and suggests directions for further research in this area.

CHAPTER 2

Detection and Analysis of Breast Masses

Methods for CAD of breast cancer in mammograms have been investigated by many researchers, resulting in a great variety of different approaches. A number of these methods are targeted at the detection of a particular sign of breast abnormality, such as the presence of calcifications, bilateral asymmetry, a dense region or mass, or architectural distortion [28]. The presence of an abnormal mass in the breast is an important sign of probable cancer. The morphological and gray-level characteristics of breast masses can be used to estimate the likelihood that a given mass is benign or malignant. Comprehensive reviews of the methods for CAD of other signs of breast cancer, which are beyond the scope of this book, have been published by Rangayyan et al. [29] and Tang et al. [30].

Algorithms for CAD of breast masses typically consist of the following tasks.

1. Locating suspicious regions (detection).

2. Extracting a region that contains a mass (segmentation).

3. Extracting and selecting features that characterize the region (analysis or feature extraction).

4. Categorizing the region as benign or malignant (classification or decision making).

The contours of breast masses presented in this book were not automatically detected and segmented but were manually drawn and verified by radiologists (see Chapter 3). Therefore, this book does not provide an in-depth review of the detection and segmentation methodologies, but instead, focuses on methodologies for the analysis and classification of breast masses based on their shape and texture or gray-scale characteristics. The following sections of this chapter provide descriptions of the characteristics of breast masses as seen on mammograms, present selected reviews of methods for CAD of breast masses, and elaborate on the methods applied in the present work for analysis of segmented breast masses.

2.1 CHARACTERISTICS OF BREAST MASSES

A mass in a breast is a 3D lesion that appears more prominent or different from the surrounding breast tissue. Breast masses may be broadly classified as benign or malignant. The likelihood of malignancy depends on the texture and morphological characteristics of the mass, such as the shape of the boundary of the mass, gray-scale variation within the mass region, and sharpness of the margins of the mass (that is, the area surrounding the mass boundary, both inside and outside the mass). Knowledge of the morphological and other characteristics that vary between benign masses and malignant tumors could allow for the development of algorithms that can accurately locate and classify them.

The shape of a mass may be described, in the order of increasing likelihood of malignancy, as round and well-circumscribed, oval, macrolobulated, microlobulated, or spiculated and ill-circumscribed [3, 31]. The gray-scale pattern within a mass region is more likely to be uniform or homogeneous for benign masses and nonuniform or heterogeneous for malignant tumors. The margin of a mass may be described, in the order of increasing likelihood of malignancy, as well-defined, obscured, or ill-defined [3, 31].

Lesions that are well-circumscribed, homogeneous, and possess well-defined boundaries are indicative of typical benign masses, whereas lesions that are spiculated, heterogeneous, and possess ill-defined boundaries are indicative of typical malignant tumors [3, 31]. (Examples of typical benign masses are shown in Figures 3.1 and 3.3, and examples of typical malignant tumors are shown in Figures 3.2 and 3.6.) There exist unusual cases of macrolobulated or slightly spiculated benign masses, as well as nearly round, microlobulated, or well-circumscribed malignant tumors; such atypical cases cause difficulties in pattern classification studies [18, 21, 25, 26]. (Figures 3.4 and 3.5 illustrate atypical benign masses and atypical malignant tumors, respectively, encountered in the dataset used in the present work.)

2.2   REVIEW OF METHODS FOR CAD OF BREAST MASSES

Algorithms for CAD of breast masses usually begin with automatic detection and segmentation of the masses, followed by extraction of features to characterize the masses, and ultimately classification of the masses as benign or malignant. This section provides a literature review of some of the methods for the diagnosis of breast masses in mammograms based on automatic segmentation approaches.

Matsubara et al. [32] applied multiple steps of thresholding to segment masses in mammograms, and classified the detected masses as benign or malignant. First, the authors employed thresholding based on histogram analysis to partition the breast into regions that contain glandular and fatty tissues, fatty tissues, and dense tissues. Next, they applied a different thresholding value to each of the breast regions segmented in the previous step to extract potential masses present in that region. Their report did not explicitly state whether the threshold values were selected manually or automatically. The potential masses were further analyzed based on their size, circularity, standard deviation, and contrast to eliminate FPs. The mass detection scheme resulted in 97% sensitivity with 3.5 FP/image. Finally, classification of the detected masses into the benign and malignant categories was achieved by examining the change in the FD of several outlines of each mass (obtained by using several different thresholds to extract the mass) and examining for the existence of spicules using the pendulum filter. The classification performance was 100% for the fractal method, and 93% sensitivity and 73% specificity for the pendulum filter method.

Mudigonda et al. [22] proposed a method to segment breast regions automatically and classify breast masses as benign or malignant. The authors applied multilevel thresholding to each mammogram to generate a map of isointensity contours, and performed grouping and merging to assign each contour to a specific group of concentric contours based on the contours' nodal relations. Each concentric group of contours represented the propagation of intensity information from the central portion of a mass region into the surrounding tissues. Thus, from this map, the authors were able to segment the image into isolated mass regions. Features based on textural flow-field information were computed from a ribbon of pixels adaptively extracted from each segmented mass margin and used to validate the segmented regions as true masses or FPs. Finally, GLCM-based texture features computed with the ribbons of successfully segmented mass regions were used to discriminate the regions as benign or malignant. The proposed methodology for segmentation resulted in 81% sensitivity with 2.2 FP/image for the discrimination of mass vs. normal tissue, and 85% sensitivity with 2.45 FP/image for the classification of malignant tumor vs. normal tissue. Classification of breast regions as benign or malignant using GLCM-based texture features resulted in an area under the ROC curve (AUC) of 0.79.

Sahiner et al. [23] developed a scheme to segment and characterize breast masses. The segmentation algorithm included three stages: K-means clustering, active contour modeling, and detection of spiculation. The K-means clustering algorithm was employed to grow and fill a mass region. The active contour model was used to refine the border of the mass. The algorithm for detection of spiculation was used to search for spicules in a band with a width of 30 pixels outside the contour of the mass. If the algorithm classified the mass as spiculated, then the algorithm combined the region encompassed by the mass outline detected by the active contour model and the region containing the spicules. After segmentation, morphological features describing the shape of the mass and texture features describing the region surrounding the mass were extracted for classification. To extract specific texture features from the region surrounding the mass, the authors designed a rubber band straightening transform (RBST) to map a band of width 40 pixels (which corresponds to 4 mm) surrounding the mass onto a rectangular region and compute texture features from the said region. A detailed description of the RBST method is provided by Sahiner et al. in an earlier publication [33]. For the task of classifying a mass on a single mammographic view, this study reported an AUC of 0.87 ± 0.02 using a combination of morphological and texture features. For the task of classifying a mass from multiple views, the study combined the scores from different views of the mass into a summary score and reported an AUC of 0.91 ± 0.02 using a combination of morphological and texture features. The results indicated that combining morphological features extracted from the automatically segmented mass boundaries with texture features improved the accuracy in CAD of mammographic masses.

Zheng and Chan [34] presented an algorithm for the detection of suspicious tumor regions in mammograms. The algorithm consists of three stages: detection, segmentation, and classification. In the detection stage, fractal analysis was applied to the given mammogram to locate regions exhibiting gray-scale or textural roughness that correspond to the presence of masses. In the segmentation stage, the discrete-wavelet-transform-based multiresolution Markov random field and the dogs-and-rabbits clustering algorithm were used to segment masses in the regions detected in the previous stage. In the classification stage, six features extracted for each segmented region were used to determine whether the region was suspicious of containing a tumor. The features selected in this study were: area, compactness, mean gradient of region boundaries, intensity variation, edge distance variation, and mean intensity difference. The proposed methodology yielded 97.3% sensitivity, 3.92 FP/image, and 0.03 false negatives (FN) per image. For future research, the authors remarked that a lower FPR could be achieved by using more features in the classification stage.

More recently, Rojas and Nandi [35] conducted a study on CAD of breast masses in mammograms with emphasis on developing robust features for automatic classification of detected mass regions. The authors identified that a limitation of most other classification algorithms lies in their strong dependency on the accuracy of the extracted mass contours. The generation of accurate mass contours requires the involvement of experts, consumes a substantial amount of time, and is not helpful in the endeavor toward CAD of breast cancer. The authors suggested that the extracted contours should only be used as pointers to the image regions from which features would be obtained. In their work, two segmentation methods, dynamic programming and constrained region growing, were separately applied to extract the mass contours. The authors wanted to observe the effect that different segmentation methods, in terms of segmentation quality, have on the effectiveness of the proposed classification features. A simplified guiding contour was produced for each extracted mass contour by fitting an ellipse to approximate the shape of the mass contour. A band or ribbon of pixels selected around the guiding contour was converted into a rectangle via the RBST. Several features designed to study the spiculation, fuzziness, contrast, and edge strength of the mass margins were computed from the RBST of the band of pixels surrounding the guiding contour. Three classifiers, Bayesian, FLDA, and support vector machine, were employed. The results indicated a 14% difference in the average segmentation quality for the constrained region growing method as compared to the dynamic programming method, but resulted in only 4% average difference in the classification performance for the constrained region growing method as compared to the dynamic programming method. It was observed that the most effective feature was the spiculation feature based on edge-signature information, which is a measure of similarity between three edge signatures (omnidirectional, radial, and tangential) of a breast mass. It was also observed in the classification experiments that malignant tumors were more difficult to predict than benign masses based on the high specificity values and low sensitivity values reported (0.8 and 0.6, respectively). The authors concluded that features extracted from automatically detected contours can contribute to the diagnosis of breast masses in screening mammograms by correctly identifying a majority of benign masses.

A few representative techniques for CAD of breast masses in mammograms were reviewed in this section. The reviewed techniques consist of automatic detection, segmentation, analysis, and classification. The ultimate objective of such methods is the realization of a CAD system to serve as a second reader to assist radiologists in their analysis of mammograms. Although several methods have demonstrated good detection sensitivity, there is still a need to reduce the FPR. The development of more sophisticated features to characterize breast masses may be the remedy to lowering the FPR of techniques for the detection and classification of masses. Methods for the detection and classification of breast masses are still under development in the effort to improve their accuracy. The following section focuses on the analysis of breast masses after they have been detected and segmented.

2.3   REVIEW OF METHODS FOR ANALYSIS OF BREAST MASSES

While a number of research teams have presented CAD systems aimed at breast masses in mammograms, others have focused on providing solutions to particular component problems in the system; in the present work, such components of interest are analysis and classification. In the analysis component, features are extracted to characterize breast masses. In the classification component, the breast masses are categorized as benign or malignant on the basis of selected features. The classification performance of features strongly depends on the accuracy of the extracted mass contours, and because contours obtained from automated segmentation methods may provide inadequate details for analysis, the present work on the analysis and classification of breast masses employs manually delineated contours; see Chapter 3 for details of the dataset used. The following subsections present details of the features employed in this book to classify breast masses based on shape and texture characteristics that vary between benign masses and malignant tumors.

2.3.1   REPRESENTATION OF BREAST MASSES FOR SHAPE AND TEXTURE ANALYSIS

For shape analysis of breast masses, the border of a breast mass is represented in two forms in the present work:

- directly as a 2D contour and

- as a one-dimensional (1D) signature derived from the 2D contour.

The 1D signature of a breast mass is obtained from its 2D contour by computing the radial (Euclidean) distance from each contour point (pixel) to the centroid of the contour as a function of the index of the contour point. A benign mass is generally round in shape, being well-circumscribed or macrolobulated, and would have a smooth signature, as shown in Figure 2.1. On the other hand, a malignant tumor is usually rough in shape, being spiculated or microlobulated, and therefore, would have a rough and complex signature, as shown in Figure 2.2.

The signature of the contour of a breast mass could also be represented as the radial distance from the centroid to the contour as a function of the angle of the radial line over the range [0°, 360°] in a polar-coordinate system [36]. However, this method leads to a multivalued function in the case of an irregular or spiculated contour (the radial lines used may intersect the contour more than once for certain angles). The signature computed in this manner would also have ranges of

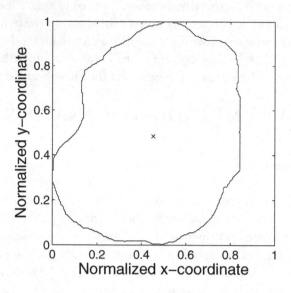

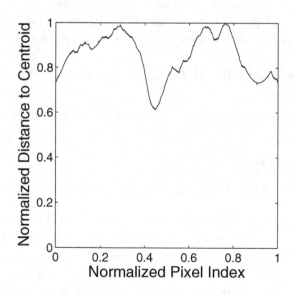

Figure 2.1: Example of the contour of a benign breast mass and the corresponding signature. The "×" mark indicates the centroid of the contour. The 2D contour and 1D signature have been normalized. Reproduced, with kind permission from Springer Science+Business Media B. V., from R. M. Rangayyan and T. M. Nguyen, "Fractal analysis of contours of breast masses in mammograms," *Journal of Digital Imaging*, 20(3):223–237, 2007. © Springer.

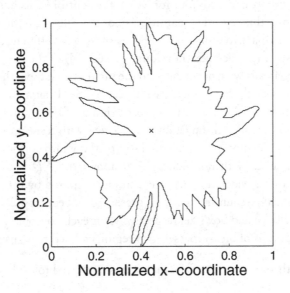

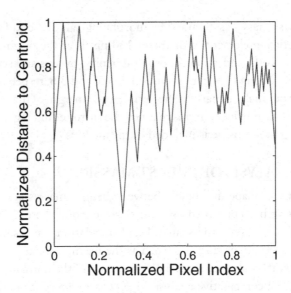

Figure 2.2: Example of the contour of a malignant breast tumor and the corresponding signature. The "×" mark indicates the centroid of the contour. The 2D contour and 1D signature have been normalized. Reproduced, with kind permission from Springer Science+Business Media B. V., from R. M. Rangayyan and T. M. Nguyen, "Fractal analysis of contours of breast masses in mammograms," *Journal of Digital Imaging*, 20(3):223–237, 2007. © Springer.

undefined values in the case of a contour for which the centroid falls outside the region enclosed by the contour. Therefore, this type of signature was not considered in the present work.

To facilitate comparative analysis of masses of widely different sizes, each 2D contour was normalized as follows: the wider axis (horizontal or vertical) of the contour was determined and all of the values along that axis were normalized to the range [0, 1]; then, the values along the other axis were normalized based on the length of the wider axis. Each 1D signature was normalized as follows: the length of the signature (that is, the range of the index of the contour points) was normalized to the range [0, 1]; then the radial distance was normalized with respect to the length of the signature. This method of normalization preserves the ratio of the width to the height of the contours and signatures in the dataset. With normalization as above, the FD values of all of the contours in the datasets, which differ widely in true size, can be computed by the ruler method (presented in Section 4.2.1) and the box-counting method (presented in Section 4.2.2) without having to change the range of the ruler size and box size, respectively, for each contour.

For texture analysis of breast masses, two regions related to a given breast mass were defined:

- the region within the border of the breast mass (referred to as the ROI) and

- the ribbon region surrounding the mass contour (referred to as the ribbon).

The ribbon of a breast mass was obtained by morphological dilation of the corresponding contour with a circular structuring element of diameter 160 pixels at the original resolution of 50 μm per pixel. (The contours, ROIs, and ribbons of several benign masses and several malignant tumors are shown in Figures 3.1, 3.3, and 3.4, and Figures 3.2, 3.5, and 3.6, respectively.) The ribbon of a mass obtained in this manner includes parts at the periphery or margin of the mass both inside and outside the boundary or contour provided; in this manner, the surrounding region of interaction of the mass with the neighboring breast tissue is included in the analysis.

2.3.2 SHAPE ANALYSIS OF BREAST MASSES

On the basis of notable shape differences between benign masses and malignant tumors, several measures of shape have been proposed for their classification. These include, but are not limited to: circularity, rectangularity, compactness, an index of spiculation, a measure of convexity or concavity, Fourier descriptors, and various statistics (such as the mean, standard deviation, entropy, area ratio, and zero-crossing count) computed from the distribution of the normalized radial length of the mass boundary. To perform a comparative analysis of FD as a feature for pattern classification of breast masses, a few selected features from the works of Rangayyan et al. [25, 26, 28] were implemented in the present work. These shape features are briefly described in the following sections.

Compactness
Compactness (C) is a measure of how efficiently a contour encloses a given area. A normalized measure of compactness is given by [37]

$$C = 1 - \frac{4\pi A}{P^2}, \tag{2.1}$$

where P and A are the contour's perimeter and area enclosed, respectively. The value of C is zero for a circle and increases with roughness or elongation of the shape. The range of C is $[0, 1]$. A high compactness value indicates a large perimeter enclosing a small area. The measure of compactness is invariant to translation, rotation, starting point, and size of the given contour, and increases in value as the shape becomes more complex and rough. Therefore, typical benign masses are expected to have lower values of compactness compared to typical malignant tumors [21, 25, 26].

Spiculation Index
Spiculation index (SI) is a measure derived by combining the ratio of the length to the base width of each possible spicule in the contour of the given mass [26]. Let S_n and θ_n, with $n = 1, 2, \ldots, N$, be the length and angle of N sets of polygonal model segments corresponding to the N spicule candidates of a mass contour. Then, SI is computed as

$$SI = \frac{\sum_{n=1}^{N} (1 + \cos\theta_n) S_n}{\sum_{n=1}^{N} S_n}. \tag{2.2}$$

The factor $(1 + \cos\theta_n)$ modulates the length of each segment (possible spicule) according to its narrowness. Spicules with narrow angles between $0°$ and $30°$ get high weighting, as compared to macrolobulations that usually form obtuse angles, and hence get low weighting. The degree of narrowness of the spicules is an important characteristic in differentiating between benign masses and malignant tumors. Benign masses are usually smooth or macrolobulated, and thus have lower values of SI as compared to malignant tumors, which are typically microlobulated or spiculated [21, 26].

Fourier Factor
The Fourier factor (FF) is a measure related to the presence of roughness or high-frequency components in a contour [37, 38]. The measure is derived by computing the sum of the normalized Fourier descriptors of the coordinates of the contour pixels divided by the corresponding indices, dividing it by the sum of the normalized Fourier descriptors, and subtracting the result from unity, as follows [37]:

$$FF = 1 - \frac{\sum_{k=-N/2+1}^{N/2} |Z_o(k)|/|k|}{\sum_{k=-N/2+1}^{N/2} |Z_o(k)|}. \tag{2.3}$$

Here, $Z_o(k)$ are the normalized Fourier descriptors, defined as

$$Z_o(k) = \begin{cases} 0, & k = 0; \\ \frac{Z(k)}{Z(1)}, & \text{otherwise.} \end{cases}$$

The Fourier descriptors themselves are defined as

$$Z(k) = \frac{1}{N} \sum_{n=0}^{N-1} z(n) \exp\left[-j \frac{2\pi}{N} nk\right], \qquad (2.4)$$

$k = -\frac{N}{2}, \ldots, -1, 0, 1, 2, \ldots, \frac{N}{2} - 1$, where $z(n) = x(n) + jy(n)$, $n = 0, 1, \ldots, N - 1$, represents the sequence of the contour's pixel coordinates. The advantages of this measure are that it is limited to the range [0, 1] and is not sensitive to noise, which would not be the case if weights increasing with frequency were used. The shape factor FF is invariant to translation, rotation, starting point, and contour size, and increases in value as the object's shape becomes more complex and rough. Contours of malignant tumors are expected to be more rough, in general, than the contours of benign masses; hence, the FF value is expected to be higher for the former than the latter [23, 25, 26].

Fractional Concavity
Fractional concavity (F_{cc}) is a measure of the portion of the indented length to the total length of the given contour. It is computed by measuring the cumulative length of the concave segments and dividing it by the total length of the contour [26]. Benign masses have fewer, if any, concave segments than malignant tumors; thus, benign masses should have lower F_{cc} values than malignant tumors [21, 26].

Examples of breast mass regions, their contours, and the corresponding values of C, SI, FF, and F_{cc} are listed in Figure 2.3. In general, it is seen that the shape features increase in value with increasing roughness of the contours.

2.3.3   TEXTURE ANALYSIS OF BREAST MASSES

Texture or gray-scale variation refers to information about the spatial arrangement of intensities in an image. Subtle textural differences have been observed between benign masses and malignant tumors, with the former being mostly homogeneous and the latter showing heterogeneous texture [3, 31]. Several studies have applied various measures of texture to discriminate between benign masses and malignant tumors [21, 22, 23, 24, 25]. The most commonly used measures of texture found in the literature were proposed by Haralick et al. [27] and are briefly described below.

Haralick's Texture Features
Haralick's texture measures [27] are statistical measures computed using the GLCM of an image. The GLCM is a square matrix with its number of rows and columns being equal to the number of gray levels in the image. For the sake of illustration, an image quantized to 3 bits per pixel (bpp),

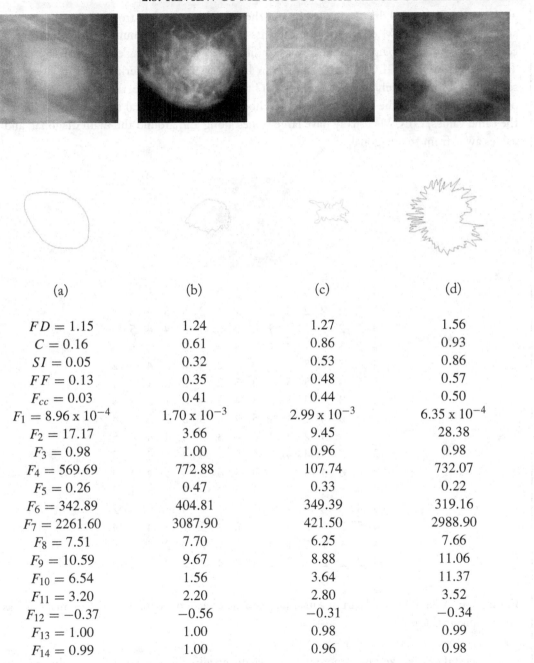

(a)	(b)	(c)	(d)
$FD = 1.15$	1.24	1.27	1.56
$C = 0.16$	0.61	0.86	0.93
$SI = 0.05$	0.32	0.53	0.86
$FF = 0.13$	0.35	0.48	0.57
$F_{cc} = 0.03$	0.41	0.44	0.50
$F_1 = 8.96 \times 10^{-4}$	1.70×10^{-3}	2.99×10^{-3}	6.35×10^{-4}
$F_2 = 17.17$	3.66	9.45	28.38
$F_3 = 0.98$	1.00	0.96	0.98
$F_4 = 569.69$	772.88	107.74	732.07
$F_5 = 0.26$	0.47	0.33	0.22
$F_6 = 342.89$	404.81	349.39	319.16
$F_7 = 2261.60$	3087.90	421.50	2988.90
$F_8 = 7.51$	7.70	6.25	7.66
$F_9 = 10.59$	9.67	8.88	11.06
$F_{10} = 6.54$	1.56	3.64	11.37
$F_{11} = 3.20$	2.20	2.80	3.52
$F_{12} = -0.37$	-0.56	-0.31	-0.34
$F_{13} = 1.00$	1.00	0.98	0.99
$F_{14} = 0.99$	1.00	0.96	0.98

Figure 2.3: Examples of (a) a circumscribed benign (CB) mass, (b) a circumscribed malignant (CM) tumor, (c) a spiculated benign (SB) mass, and (d) a spiculated malignant (SM) tumor, and the values of their shapes and texture measures.

as shown in Figure 2.4, has 8 gray levels, and the GLCM computed from the image is a matrix of size 8×8. The GLCM of an image represents a tabulation of the numbers of occurrences of all combinations of pixel pairs, separated by a specified distance, d, and angle, θ, occurring in the image. Table 2.1 shows the GLCM for the image in Figure 2.4 by considering pairs of pixels with the second pixel immediately next to (the right of) the first. For example, the pair of gray levels [1 2] occurs eight times in the image. Because neighboring pixels in natural images tend to have nearly the same values, GLCMs usually have large values along and around the main diagonal, and small values away from the diagonal.

```
4 1 1 2 0 0 3 2 2 2 3 3 3 5 3 4
3 2 2 2 2 3 2 2 2 1 2 5 3 4 3 4
3 1 2 1 1 4 5 4 2 2 3 5 4 4 5 5
4 2 3 3 1 3 4 5 4 4 3 5 5 6 6 3
4 5 5 5 3 2 3 4 3 2 4 4 3 4 4 2
5 5 3 5 5 5 5 3 3 2 3 3 2 1 0 1
4 4 3 3 3 5 6 5 4 5 4 3 1 1 2 4
2 2 3 4 2 4 4 5 4 4 3 3 2 3 4 4
2 3 5 5 2 4 4 6 4 3 3 3 4 4 4 5
3 5 7 5 4 4 2 6 5 5 4 4 4 4 4 3
5 6 6 4 4 5 3 4 5 4 4 3 5 4 1 3
5 4 4 4 4 5 4 2 3 4 4 5 5 5 1 3
5 4 3 2 4 4 4 1 3 4 4 4 5 5 4 2
4 2 2 1 2 3 3 3 2 2 2 1 2 4 5 4
4 2 1 1 1 1 0 2 3 1 2 3 2 3 5 5
3 2 2 2 2 3 2 3 4 4 2 2 2 1 2 2
```

Figure 2.4: A 16×16-pixel part of a mammogram quantized to 3 bpp, shown as an image and as a 2D array of pixel values.

Digital or digitized mammograms are usually acquired at the resolution of about 50 μm per pixel, with 4096 gray levels represented using 12 bpp. However, the 4096×4096 GLCM of a 12-bpp image would be excessively large for practical computation; furthermore, most pairs of gray levels would occur with low or negligible rates of incidence that may not permit the derivation of reliable

Table 2.1: Gray-level cooccurrence matrix for the image in Figure 2.4, with the second pixel immediately next to (the right of) the first. Pixels in the last column were not processed as current pixels. The GLCM has not been normalized.

Current Pixel	Next Pixel							
	0	1	2	3	4	5	6	7
0	1	1	1	1	0	0	0	0
1	2	6	8	4	1	0	0	0
2	1	7	18	16	6	1	1	0
3	0	4	13	12	13	9	0	0
4	0	3	10	11	27	13	1	0
5	0	1	1	6	15	14	3	1
6	0	0	0	1	2	2	2	0
7	0	0	0	0	0	1	0	0

statistics. Therefore, in order to avoid sparse GLCMs, it is advantageous to reduce the image to 256 gray levels, that is, to use 8 bpp. The images in the datasets used in the present work are quantized at 8 bpp. The normalized GLCM is computed as:

$$p_{(d,\,\theta)}(l_1, l_2) = \frac{n_{l_1 l_2}}{\sum_{l_1=0}^{L-1} \sum_{l_2=0}^{L-1} n_{l_1 l_2}}, \tag{2.5}$$

where L is the total number of gray levels in the image; l_1 and l_2, with each variable in the range $0, 1, 2, \ldots, L - 1$, represent a pair of gray levels; $n_{l_1 l_2}$ is the number of occurrences of the pair of gray levels l_1 and l_2; and $p_{(d,\,\theta)}(l_1, l_2)$ is the probability of occurrence of the pair of gray levels l_1 and l_2 at the specified spatial distance d and angle θ.

A GLCM is often observed to have high values concentrated along and around its diagonal, and low values elsewhere in the rest of the entries of its matrix. This is due to the fact that neighboring pixels in an image tend to have similar intensity. The GLCM is often computed for unit spatial distance, $d = 1$, and four different angles, $\theta = 0°, 45°, 90°$, and $135°$. A unique GLCM may be computed for each pair of d and θ values to observe the dependence of texture upon angle, or a single GLCM, $p(l_1, l_2)$, may be computed as the average of the GLCMs computed for each pair of d and θ for all angles ($\theta = 0°, 45°, 90°,$ and $135°$) if the dependence of texture upon angle is not of interest. The distance d should be chosen with consideration of the size (or resolution) of the objects of interest or scale of variation present in the image. The choice of d determines the resolution at which the texture is analyzed. Thus, one of the objectives of the present study is to analyze the effect of spatial resolution or pixel size on GLCM-based texture features in the classification of breast masses.

Fourteen statistical measures were derived from GLCMs by Haralick et al. [27]. These measures or features reveal properties about the spatial distribution of gray levels in the image. All of the 14 features are briefly described in the following paragraphs.

- Angular second moment: A measure of homogeneity or uniformity in the image. A homogeneous image has similar pixel intensity throughout; thus, the GLCM of the image will have a small number of high values concentrated along the diagonal. On the other hand, the GLCM of an inhomogeneous image will have small values spread over a large number of entries of its matrix. The angular second moment is defined as

$$F_1 = \sum_{l_1=0}^{L-1} \sum_{l_2=0}^{L-1} p^2(l_1, l_2). \tag{2.6}$$

- Contrast: A measure of the amount of local variations present in the image. Contrast is defined as

$$F_2 = \sum_{k=0}^{L-1} k^2 \underbrace{\sum_{l_1=0}^{L-1} \sum_{l_2=0}^{L-1} p(l_1, l_2)}_{|l_1 - l_2| = k}. \tag{2.7}$$

- Correlation: A measure of linear dependencies of gray levels in the image (that is, how a pixel is correlated to its neighborhood). The correlation measure is defined as

$$F_3 = \frac{1}{\sigma_x \sigma_y} \left[\sum_{l_1=0}^{L-1} \sum_{l_2=0}^{L-1} l_1 \, l_2 \, p(l_1, l_2) - \mu_x \, \mu_y \right], \tag{2.8}$$

where μ_x and μ_y are the means, and σ_x and σ_y are the standard deviation values of p_x and p_y, respectively. The marginal probabilities, p_x and p_y, are defined as

$$p_x(l_1) = \sum_{l_2=0}^{L-1} p(l_1, l_2), \qquad (2.9)$$

$$p_y(l_2) = \sum_{l_1=0}^{L-1} p(l_1, l_2). \qquad (2.10)$$

Then, we have

$$\mu_x = \sum_{l_1=0}^{L-1} l_1 \, p_x(l_1) \qquad (2.11)$$

and

$$\sigma_x^2 = \sum_{l_1=0}^{L-1} (l_1 - \mu_x)^2 \, p_x(l_1) \qquad (2.12)$$

with similar definitions for μ_y and σ_y using p_y.

- Sum of squares: A measure of the gray-level variance in the image.

$$F_4 = \sum_{l_1=0}^{L-1} \sum_{l_2=0}^{L-1} (l_1 - \mu_f)^2 \, p(l_1, l_2), \qquad (2.13)$$

where μ_f is the mean gray level of the image.

- Inverse difference moment: A measure of local uniformity present in the image. The inverse difference moment is defined as

$$F_5 = \sum_{l_1=0}^{L-1} \sum_{l_2=0}^{L-1} \frac{1}{1 + (l_1 - l_2)^2} \, p(l_1, l_2). \qquad (2.14)$$

This feature may be seen as the inverse of the contrast feature defined in Equation 2.7. Thus, the inverse difference moment is high for images having low contrast and low for images having high contrast.

- Sum average: The sum average is defined as

$$F_6 = \sum_{k=0}^{2(L-1)} k \, p_{x+y}(k), \qquad (2.15)$$

where p_{x+y} is given by

$$p_{x+y}(k) = \underbrace{\sum_{l_1=0}^{L-1} \sum_{l_2=0}^{L-1}}_{l_1+l_2=k} p(l_1, l_2).$$

(2.16)

• Sum variance: The sum variance is defined as

$$F_7 = \sum_{k=0}^{2(L-1)} (k - F_6)^2 \, p_{x+y}(k).$$

(2.17)

• Sum entropy: The sum entropy is defined as

$$F_8 = -\sum_{k=0}^{2(L-1)} p_{x+y}(k) \, \log_2 \left[p_{x+y}(k) \right].$$

(2.18)

• Entropy: A measure of the nonuniformity or complexity of the texture of the image. The entropy is defined as

$$F_9 = -\sum_{l_1=0}^{L-1} \sum_{l_2=0}^{L-1} p(l_1, l_2) \, \log_2 \left[p(l_1, l_2) \right].$$

(2.19)

• Difference variance: The difference variance is computed, in a manner similar to the sum variance, for the p_{x-y} matrix, as

$$F_{10} = \sum_{k=0}^{2(L-1)} \left(k - \sum_{k=0}^{2(L-1)} k \, p_{x-y}(k) \right)^2 p_{x-y}(k),$$

(2.20)

where p_{x-y} is given by

$$p_{x-y}(k) = \underbrace{\sum_{l_1=0}^{L-1} \sum_{l_2=0}^{L-1}}_{|l_1-l_2|=k} p(l_1, l_2).$$

(2.21)

- Difference entropy: The difference entropy is defined as

$$F_{11} = -\sum_{k=0}^{L-1} p_{x-y}(k) \, \log_2 \left[p_{x-y}(k) \right] . \tag{2.22}$$

- Two information measures of correlation: The information measures of correlation are defined as

$$F_{12} = \frac{H_{xy} - H_{xy1}}{\max\{H_x, H_y\}} , \tag{2.23}$$

and

$$F_{13} = \left\{ 1 - \exp[-2 \, (H_{xy2} - H_{xy})] \right\}^{\frac{1}{2}} , \tag{2.24}$$

where $H_{xy} = F_9$; H_x and H_y are the entropies of p_x and p_y, respectively;

$$H_x = -\sum_{l_1=0}^{L-1} p_x(l_1) \, \log_2 \left[p_x(l_1) \right] , \tag{2.25}$$

$$H_{xy1} = -\sum_{l_1=0}^{L-1} \sum_{l_2=0}^{L-1} p(l_1, l_2) \, \log_2 \left[p_x(l_1) \, p_y(l_2) \right] , \tag{2.26}$$

and

$$H_{xy2} = -\sum_{l_1=0}^{L-1} \sum_{l_2=0}^{L-1} p_x(l_1) \, p_y(l_2) \, \log_2 \left[p_x(l_1) \, p_y(l_2) \right] . \tag{2.27}$$

- Maximum correlation coefficient: The maximum correlation coefficient is defined as

$$F_{14} = (Second \; largest \; eigenvalue \; of \; Q)^{1/2} , \tag{2.28}$$

where Q is computed as

$$Q(l_1, l_2) = \sum_{k=0}^{L-1} \frac{p(l_1, k) \, p(l_2, k)}{p_x(l_1) \, p_y(k)} . \tag{2.29}$$

Further details of the characteristics of the 14 statistical measures of texture based upon the GLCM are given by Haralick et al. [27]. These measures represent characteristics of the spatial distribution of gray levels in the image. Texture analysis, using some or all of Haralick's 14 texture features, is a popular approach for the analysis and classification of many medical images, including breast masses and tumors seen in mammograms [21, 22, 23, 24, 25, 28]. Therefore, in the present work, Haralick's texture measures have been chosen for the analysis of the texture of breast masses.

Examples of breast mass regions, the contours used to extract the related ROIs, and the corresponding values of Haralick's texture measures are listed in Figure 2.3. It is difficult to identify simple trends in the values of the texture features.

2.4   REMARKS

This chapter presented a discussion on the shape and gray-scale characteristics that vary between benign masses and malignant tumors in mammograms, provided a brief review of the methods for CAD of breast masses reported in the literature, and elaborated on the methods applied in the present work for the analysis of shape and texture of breast masses. Shape and texture features extracted from mass regions are often sensitive to the exact shape of the contours used to extract the related ROIs. Therefore, the contours of breast masses used in this book were delineated manually by expert radiologists. The shape features C, SI, FF, and F_{cc} described in this chapter are used in comparative analysis with FD as shape features for the classification of breast masses. Haralick's texture measures, also described in this chapter, are computed for both the ROIs and the ribbons of pixels around the masses. A comparison of the classification performance of Haralick's texture measures computed with ROIs vs. the same measures computed with ribbons of masses is presented in Chapter 6. Furthermore, the effect of spatial resolution or pixel size on Haralick's texture measures in the classification of breast masses is also discussed in Chapter 6. This chapter has established the necessary background of the methods for the analysis of shape and texture of breast masses used in this book.

CHAPTER 3

Datasets of Images of Breast Masses

Three datasets of images of breast masses were used in this study. The images for the first dataset were obtained from Screen Test: the Alberta Program for the Early Detection of Breast Cancer [11, 21, 39]. The mammograms in this dataset are from 20 cases. The mammograms were digitized using the Lumiscan 85 scanner at a resolution of 50 μm with 12 bpp. The dataset prepared for this study includes 57 ROIs, 37 of which are related to benign masses and 20 are related to malignant tumors. The areas of the benign masses vary in the range 39–423 mm^2, with an average of 163 mm^2 and a standard deviation of 87 mm^2. The areas of the malignant tumors vary in the range 34–1,122 mm^2, with an average of 265 mm^2 and a standard deviation of 283 mm^2. The diagnostic classification of the masses was based upon biopsy. The contour of each mass was manually drawn by an expert radiologist specialized in mammography and verified independently by a second radiologist. A ribbon of pixels around each mass was extracted by dilating the contour with a structuring element of diameter 8 mm and applying the result as a mask on the original ROI. Most of the benign masses in this dataset are well-circumscribed (CB), whereas most of the malignant tumors are spiculated (SM), as typically encountered in mammographic images. Figures 3.1 and 3.2 illustrate the contours, ROIs, and ribbons of two benign masses and two malignant tumors, respectively, from the first dataset.

The second dataset was obtained by using images containing masses from the Mammographic Image Analysis Society (MIAS, UK) database [40, 41] and the teaching library of the Foothills Hospital in Calgary [25, 26]. The MIAS images were digitized at a resolution of 50 μm with 8 bpp, whereas the Foothills Hospital images were digitized at a resolution of 62 μm with 8 bpp. The diagnostic classification of the masses was based upon biopsy. The contour of each mass was manually drawn by an expert radiologist specialized in mammography. This dataset includes circumscribed and spiculated cases in both the benign and malignant categories. SB masses and CM tumors are unusual, and tend to cause difficulties in pattern classification studies [25, 26]. The second dataset has a total of 54 masses, including 16 CB, 12 SB, 19 SM, and 7 CM types. The areas of the benign masses vary in the range 32–1,207 mm^2, with an average of 281 mm^2 and a standard deviation of 288 mm^2. The areas of the malignant tumors vary in the range 46–1,244 mm^2, with an average of 286 mm^2 and a standard deviation of 292 mm^2. The contours, ROIs, and ribbons of a few examples of circumscribed and spiculated cases of benign masses and malignant tumors from the second dataset are illustrated in Figures 3.3–3.6.

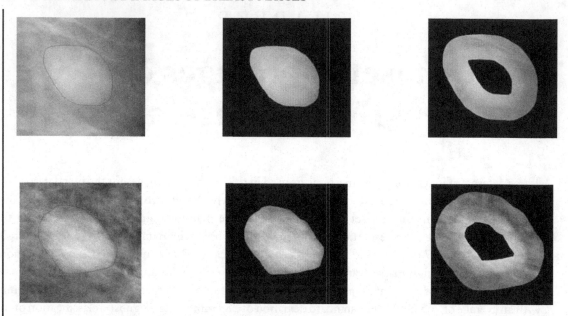

Figure 3.1: Illustrations of two CB masses from the first dataset, where each row represents the contour superimposed on a part of the original mammogram, ROI, and ribbon of a mass.

Figure 3.2: Illustrations of two SM tumors from the first dataset, where each row represents the contour superimposed on a part of the original mammogram, ROI, and ribbon of a tumor.

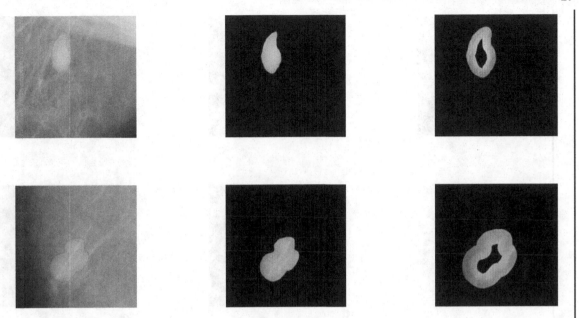

Figure 3.3: Illustrations of two CB masses from the second dataset, where each row represents the contour superimposed on a part of the original mammogram, ROI, and ribbon of a mass.

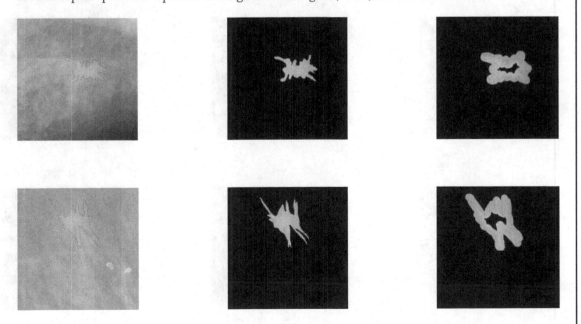

Figure 3.4: Illustrations of two SB masses from the second dataset, where each row represents the contour superimposed on a part of the original mammogram, ROI, and ribbon of a mass.

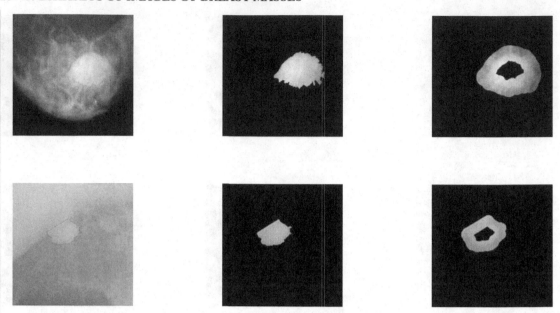

Figure 3.5: Illustrations of two CM tumors from the second dataset, where each row represents the contour superimposed on a part of the original mammogram, ROI, and ribbon of a tumor.

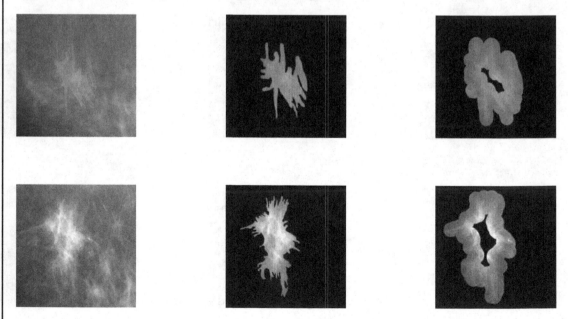

Figure 3.6: Illustrations of two SM tumors from the second dataset, where each row represents the contour superimposed on a part of the original mammogram, ROI, and ribbon of a tumor.

The third (combined) dataset was prepared by combining all of the cases in the first and the second datasets. The areas of the benign masses in the combined dataset vary in the range $32–1,207$ mm^2, with an average of 214 mm^2 and a standard deviation of 206 mm^2. The areas of the malignant tumors in the combined dataset vary in the range $34–1,244$ mm^2, with an average of 277 mm^2 and a standard deviation of 285 mm^2. The combined set has 111 contours including both typical and atypical shapes of benign masses (65) and malignant tumors (46). Regardless of minor variations in the imaging and data acquisition protocols, the combined dataset includes a good number of masses of different types and has been used in several studies on CAD [18, 19, 21, 22, 24, 25, 26].

The classification results obtained in the present study are presented for the three datasets (first and second datasets separately and combined) in order to analyze and demonstrate the strengths and weaknesses of the features used in characterizing breast masses and tumors of various types (CB, SB, SM, and CM).

CHAPTER 4

Methods for Fractal Analysis

Fractal geometry has received considerable attention as a mathematical tool to model randomness in nature. In signal and image processing, fractal analysis is widely used to characterize the shape complexity and to assess the texture or roughness of objects. The present work examines the application of fractals to classify breast masses based on the irregularity exhibited in their contours and the gray-scale variability exhibited in their images. The first section of this chapter presents the theory of fractals, provides examples of well-known fractal patterns, gives examples of the presence of fractal geometry in biology, and gives a review of related research works that have made use of fractals. The second section of this chapter provides descriptions of methods to compute the FD of contours of breast masses based on their irregularity. The third section of this chapter presents methods to compute the FD of breast mass regions based on their gray-scale variability.

4.1 FRACTALS

The term "fractal" was coined by Mandelbrot [20] in the mid-1970s to label functions or patterns that possess self-similarity at all scales or levels of magnification [42, 43, 44, 45, 46, 47]. The notion of self-similarity refers to the characteristic that an object exhibits when its substructure resembles its superstructure in the same form. This means that, under all magnification levels, a fractal appears identical to its original (unmagnified) form. The quality of being self-similar at all scales differentiates fractals from traditional Euclidean shapes (for example, a circle, a triangle, and a square). A Euclidean shape appears less and less like itself as it is magnified. At infinite magnification, it would be impossible to differentiate between any two Euclidean shapes.

Although the term fractal, used to describe objects exhibiting self-similarity, is considered to be relatively new in the history of mathematics, there is substantial evidence that the idea of self-similarity has been in existence since the 1800s. Such evidence is found in the Cantor bar, Koch curve, Sierpinski triangle, and Hilbert curve, which are all considered as well-known or famous fractals [42]. The processes to generate these fractals are reviewed in Section 4.1.1.

A fractal is usually constructed using a recursive algorithm, whereas a Euclidean shape, such as a circle, is constructed by an algebraic formula. For example, the equation or formula $r^2 = x^2 + y^2$ defines a circle with radius r located at the origin of the (x, y) plane. Fractals, being generated in a recursive manner, possess irregular and highly complex appearances that cannot be described adequately using combinations of simple Euclidean shapes. Similarly, the dimension of a fractal cannot be described adequately by using the traditional Euclidean dimension or topological dimension; rather, a measure such as FD is used to lead to a quantitative representation of the complex shapes and forms of fractals. The self-similarity dimension (discussed in Section 4.1.3), compass dimension (discussed in Section 4.2.1), and box-counting dimension (discussed in Section 4.2.2) are all considered to be special forms of Mandelbrot's FD [42, 43, 48].

Most objects in nature do not resemble Euclidean shapes. According to Mandelbrot [20], "Clouds are not spheres, mountains are not cones, coastlines are not circles, and bark is not smooth, nor does lightning travel in a straight line." Hence, fractal geometry is used to describe, model, and analyze the seemingly complex (chaotic and random) forms found in nature. This is not to say that naturally occurring objects possess strict self-similarity at all scales similar to that of ideal fractals. Firstly, an object in nature can never have infinite magnification because it cannot possess infinite granularity. Secondly, an object in nature, when magnified, appears similar, but never exactly identical, to its original (unmagnified) form. This characteristic is referred to as statistical self-similarity. A commonly cited example for demonstrating the notion of statistical self-similarity is through the task of examining the coastline of Britain under a range of magnification. Upon magnification, segments of the coastline appear similar to, but never exactly like, segments at different scales. Therefore, naturally occurring objects are approximate fractals which display statistical self-similarity over a finite range of scales. Further discussion on the presence of fractal geometry in nature is presented in Section 4.1.2.

Often times, objects and patterns found in nature exhibit a high degree of randomness. For such cases, a popular mathematical model, called fractional Brownian motion (fBm) [42, 46, 48, 49], can be used to model the randomness. The fBm model is an extension of Brownian motion, which is also known as the "random walk process." Brownian motion, named after Robert Brown, was introduced to model the random movement of a microscopic particle, suspended in a liquid or gas, as a result of the cumulative effect of the collisions with the surrounding particles in the medium. A Brownian motion curve representing the random motion of a particle in only one space variable with respect to time is explained as follows. At time $t = 0$, the position of the particle, $V(t)$, is $V(0) = 0$. At time $t = 1$, the particle displaces by a certain length l in either direction (with both directions having equal probability of 0.5). The displacement l is a random variable with a Gaussian distribution. Furthermore, successive displacements of the particle are independent of one another. The total displacement of the particle after a certain time interval is zero, but the mean-squared displacement is proportional to the time difference. In other words, for ordinary or standard Brownian motion, the increment $V(t_2) - V(t_1)$ has a Gaussian distribution with variance

$$var[V(t_2) - V(t_1)] \propto |t_2 - t_1|^{2H},\qquad(4.1)$$

where $H = 1/2$ and is called the Hurst exponent [42].

The 1D Brownian motion described above can be examined for self-similarity by scaling its time and amplitude axes. It was discovered that scaling Brownian motion in time by a factor of 2 and in amplitude by a factor of $\sqrt{2}$ resulted in a curve that appears similar to the original, unscaled, Brownian motion curve [42]. The exponent H describes the scaling factor and is called the Hurst exponent [42]. A case of Brownian motion with $H = 1/2$ is referred to as standard Brownian motion, whereas Brownian motion with $0 < H < 1$ is known as fBm. For $1/2 < H < 1$, there is a positive correlation between time increments; this means that, if there is an increasing pattern in the previous step, then it is likely that the current step will be increasing as well. For $0 < H < 1/2$, there is a negative correlation between time increments; the opposite of the effect of positive correlation as described above is true and the fBm pattern for negative correlation will appear to oscillate more erratically. The nonuniform scaling property of fBm is known as self-affinity. The fBm process is often used to model randomness in naturally occurring objects such as clouds and mountains.

In general, any varying quantity, $V(t)$, with the best-fitting line to its power spectral density (PSD), $P_V(f)$, varying as $1/f^\beta$ on a log–log scale, is referred to as $1/f$ noise. The parameter β is known as the spectral component and has been observed to have values in the range $[0.5, 1.5]$ for most natural phenomena [48]. This relationship is known as the $1/f$ model or the inverse power law.

Figure 4.1 shows examples of signals generated as functions of time (sample number), $V_H(t)$, based on the fBm model [50]. The scaling of the traces is characterized by the scaling parameter H in the range $0 \le H \le 1$. A high value of H (close to 1) gives a smooth trace, whereas a low value of H produces a rough trace. The parameter H (the Hurst coefficient) relates the changes in V, $\Delta V = V(t_2) - V(t_1)$, to the time difference, $\Delta t = t_2 - t_1$, by the scaling law expressed as [48] $\Delta V \propto (\Delta t)^H$.

Self-similar patterns repeat themselves under magnification; however, fBm traces repeat statistically only when t and V are magnified by different amounts [50]. If t is magnified by a factor r as rt, the value of V is magnified by the factor r^H and becomes $r^H V$. This form of nonuniform scaling is known as self-affinity [20]. The zero-set of an fBm signal is the intersection of the signal with the horizontal axis. The zero-set is a set of disconnected points with a topological dimension of zero and FD, expressed here as D_0, given by [48] $D_0 = 1 - H$. The zero-set of a self-affine signal is self-similar; different estimates of D_0 will yield the same result. The FD of the original signal is related to D_0 as $FD = D_0 + 1$, and to the scaling parameter, H, as $FD = 2 - H$.

Methods to compute FD of signals and images exhibiting fBm characteristics are presented in Sections 4.2.3 and 4.3.2, respectively.

4.1.1 FAMOUS FRACTALS

The Cantor bar, Koch curve, and Sierpinski triangle are three commonly recognized fractals; these patterns can be generated by repeating a basic pattern in a recursive or iterative process [42, 43]. The Cantor bar, shown in Figure 4.2, is the most simple of the three. It is formed by repeatedly dividing a line segment into three equal segments and eliminating the middle segment. Every successive iteration causes the length of the Cantor bar to be 2/3 of its previous length. After a number of iterations, the Cantor bar appears like a collection of disconnected points. After an infinite number of iterations, the length becomes zero.

The Koch curve, shown in Figure 4.3, is formed by dividing a line segment into three equal segments, then replacing the middle segment with an equilateral triangle whose base is removed.

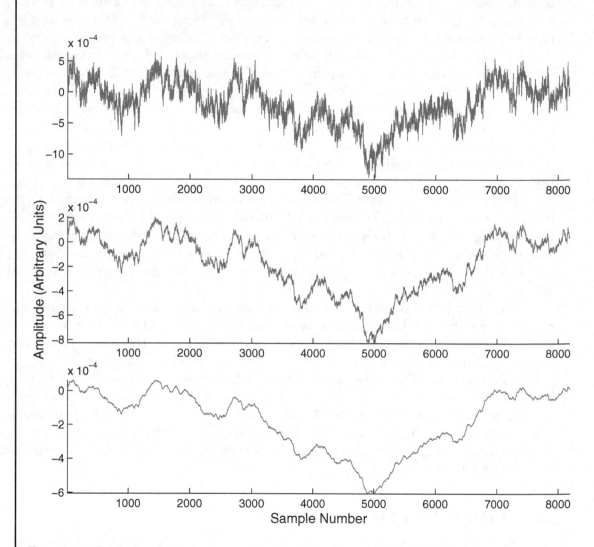

Figure 4.1: Examples of signals generated based on the fBm model for different values of H and FD. Top to bottom: $H = 0.2, 0.5,$ and 0.8; model $FD = 1.8, 1.5,$ and 1.2; estimated $FD = 1.8070, 1.5076,$ and 1.2081. Reproduced with permission from R. M. Rangayyan and F. Oloumi, "Fractal analysis and classification of breast masses using the power spectra of signatures of contours," *Journal of Electronic Imaging*, 21(2):023018, 2012. © SPIE.

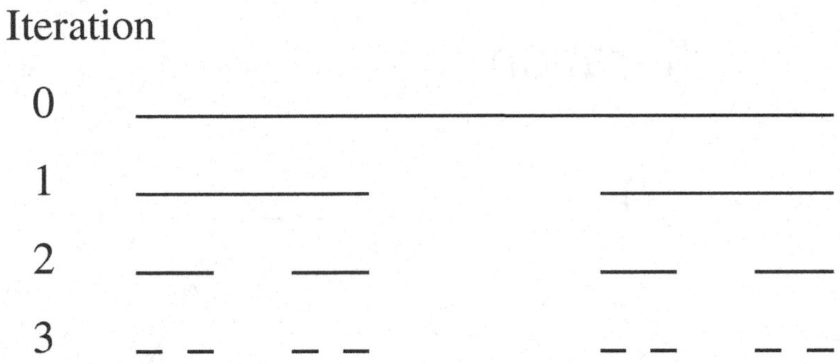

Iteration

0 _____

1 _____   _____

2 ____   ____   ____   ____

3 __ __   __ __   __ __   __ __

Figure 4.2: An illustration of the initial state and the first three iterations of the construction of a Cantor bar.

The same process is repeated with each new line segment. After an infinite number of iterations, the length of the Koch curve becomes infinite. An extension of the Koch curve is the Koch snowflake, which is formed in the same manner but starting with an equilateral triangle rather than a line segment (see Figure 4.14). The perimeter of the Koch snowflake increases by 4/3 of the previous perimeter for each iteration and tends toward infinity.

The Sierpinski triangle, shown in Figure 4.4, is constructed by beginning with, usually, but not necessarily, an equilateral triangle, then connecting the midpoints of each side of the triangle to form four smaller triangles with the middle triangle removed. The same process is performed on each newly generated triangle.

Fractals may also be described in terms of their space-filling nature. The Hilbert curve [42] is one of the well-recognized fractals that demonstrates this behavior. The intuitive understanding of dimension is that a curve is 1D and a plane is 2D. In contrast to this notion of dimension, Hilbert demonstrated that a curve can fill a 2D plane, that is, after an infinite number of iterations, the Hilbert curve will pass through every point in a plane without crossing itself. The construction of the Hilbert curve is explained through Figure 4.5. In the initial state, the Hilbert curve connects the center points of four subsquares as shown in the first case in Figure 4.5. In the first iteration, each of the four squares is divided into four more squares to create a new grid, and the original curve (or curve from the previous iteration) is reduced to half of its size. For the third and fourth quadrants of the new grid, the curve from the previous iteration is placed in the same orientation. For the first quadrant of the new grid, the curve from the previous iteration is rotated 90° clockwise and placed. For the second quadrant of the new grid, the curve from the previous iteration is rotated 90° counterclockwise and placed. The curves from the four quadrants of the grid are connected together with three line segments to form a new curve. The same process is repeated for the next iteration with the newly generated curve. In every iteration, the Hilbert curve traces through the center of every subsquare of the grid in a continuous manner without crossing itself. Eventually, the Hilbert curve will appear to have filled the entire grid at finer and finer resolution. An important property

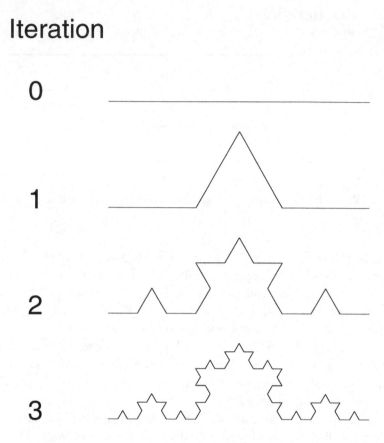

Figure 4.3: An illustration of the initial state and the first three iterations of the construction of a Koch curve.

of the Hilbert curve, also referred to as the Peano curve or the Peano-Hilbert curve [51], is that it visits all of the subsquares or their center points in a given quadrant or subquadrant of the given space before leaving the same for the next subquadrant.

4.1.2   FRACTALS IN BIOLOGY

The concept of fractals has changed the way researchers perceive and characterize the shapes and patterns of objects found in nature. The structures of some objects in nature, such as ferns, broccoli, cauliflower, and tree branches, have elements of similarity and randomness over a range of scale. These objects are said to exhibit fractal geometry.

A number of anatomical structures demonstrate fractal patterns, such as the coronary arteries, retinal vasculature [52, 53, 54, 55], venous branching patterns, bronchial trees, certain muscle-fiber

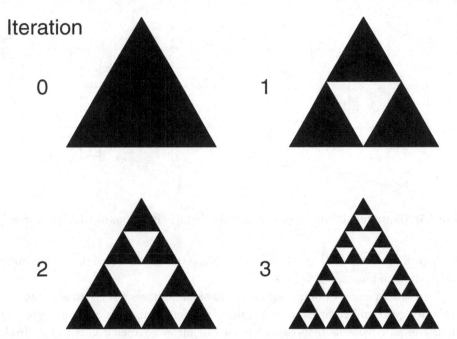

Figure 4.4: An illustration of the initial state and the first three iterations of the construction of a Sierpinski triangle.

bundles, and the His-Purkinje network in the ventricles [44, 56]. The His-Purkinje network of conduction pathways provides an efficient way to distribute the depolarization stimulus to the ventricles. The electrogenesis of the QRS complex in the electrocardiogram (ECG) has been modeled using a fractal-like conduction system. The normal QRS complex has been shown to have an inverse-power-law distribution of frequency content in the log–log scale [57], which agrees with depolarization of the myocardium by a self-similar branching network [56]. Computer modeling studies of self-similar branching networks used to depolarize a network of cells have shown that, after 10 generations of branching, it is possible to simulate realistic QRS complexes [58]. Changes in the geometry of the branching network have been shown to affect the frequency content of the associated QRS complexes.

Goldberger et al. [57] and Goldberger and West [59] showed that the healthy heart beat is a temporal fractal that is not highly regular, but instead, has a high degree of variability in time series of heart rate. The PSD of a time-series representation of heart rate has a $1/f$-like distribution and follows the inverse power law. It has been shown that, in the case of time series of heart rate, the loss of physiological complexity can lead to greater regularity. The concepts of fractals, self-similar scaling, $1/f$ noise, and inverse-power-law distributions offer new ways to describe anatomical structures and

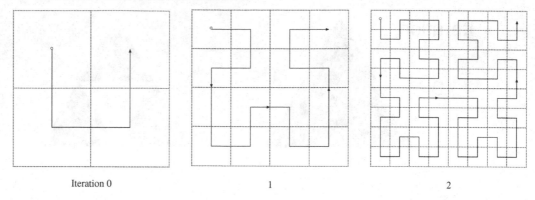

Iteration 0   1   2

Figure 4.5: Illustration of the initial state and the first two iterations of a Hilbert curve.

physiological processes [59]. See Rangayyan et al. [60] for additional discussion on applications of fractal analysis to biomedical signals.

Cancerous tumors exhibit a certain degree of randomness associated with their growth, and are typically irregular and complex in shape; therefore, fractal analysis could provide a better measure of their complex patterns than conventional Euclidean geometry. Gazit et al. [61] showed that the vascular architecture of tumors during growth displays fractal characteristics that significantly differ from those of normal vascular networks of healthy tissues. It was observed that the FD of the vascular architecture of growing tumors is significantly lower than that of normal vascular architecture, with the latter demonstrating a clear space-filling nature. It was also observed that, during tumor regression, the FD values of the vasculature are in between those of growing tumors and healthy tissues. It is not known if a direct relationship exists between the internal vascular architecture and the external shape or border characteristics of a tumor. Nevertheless, the complex and rough patterns of the contours of malignant tumors can be characterized by FD, and they may be expected to have higher FD values than the relatively smooth contours of benign masses [18].

4.1.3   FRACTAL DIMENSION

The shapes of objects and organisms have traditionally been described using Euclidean geometry. This type of geometry defines the dimension of a line as one, a flat surface (such as a square) as two, and solids (such as a cube) as three. However, objects found in nature can rarely be described using basic geometrical shapes. Consider the example of the leaf of a fern, shown in Figure 4.6, which exhibits fractal characteristics; it is clearly inadequate to describe the shape of such a leaf as being composed of triangular or oval sections. The details and complexity of irregular shapes are neglected in the traditional sense of dimension. The notion of FD brings about a revision in the way dimension is defined, such that a jagged line does not have the same dimension as a straight line, but a dimension falling in the range of 1–2, depending on the degree or extent of its jaggedness. In this sense, FD provides a quantitative measure of the complexity of a function or an object.

Figure 4.6: A fern leaf exhibiting fractal patterns. Reproduced with permission from R. M. Rangayyan, *Biomedical Image Analysis*, CRC Press, Boca Raton, FL, 2005. © CRC Press.

The self-similarity dimension (D) is defined as follows [42]. Consider a self-similar pattern that exhibits a number of self-similar pieces at the reduction factor of $1/s$ (the latter is related to the measurement scale). The power law expected to be satisfied is

$$a = \frac{1}{s^D}. \tag{4.2}$$

Taking the log of both sides of the above equation and solving for D gives

$$D = \frac{\log(a)}{\log(1/s)}. \tag{4.3}$$

Therefore, the slope (of the straight-line approximation) of a plot of the log of the number of self-similar pieces, $\log(a)$, vs. the log of the reduction factor, $\log(1/s)$, henceforth referred to as the log–log curve, can provide an estimate of D, which is the self-similar dimension.

Consider, for example, the Koch curve shown in Figure 4.3. The Koch curve begins from a line; then, the first iteration generates four lines or self-similar pieces that are each $1/3$ the size of the original line. Therefore, the self-similar dimension of the Koch curve is $\log(4)/\log(3) = 1.26$.

4.1.4   APPLICATIONS OF FRACTAL ANALYSIS

The notion of fractal analysis [20, 42, 43, 44, 45, 46, 47, 48] is useful in studying the complexity of 1D functions, 2D contours, as well as gray-scale images. A few studies have examined the application of fractals to classify breast masses based on the irregularity exhibited in their contours. A study by Matsubara et al. [32] reported 100% accuracy in the classification of 13 benign masses and malignant tumors using FD. The method employed by Matsubara et al. involved the computation of a series of FD values for several contours of a given mass obtained by thresholding the mass at many levels; the change in FD of the given mass was used to categorize the mass as benign or malignant. A study by Pohlman et al. [36] obtained higher than 80% classification accuracy with fractal analysis of signatures of contours of breast masses. However, the signature of a contour was derived as a function of the radial distance from the centroid to the contour vs. the angle of the radial line over the range $[0°, 360°]$, which could lead to a multivalued function in the case of an irregular or spiculated contour; the signature computed in this manner would also have ranges of undefined values in the case of a contour for which the centroid falls outside the region enclosed by the contour. Dey and Mohanty [62] employed fractal geometry to study breast lesions on cytology smears and found that FD may be useful in discriminating between benign and malignant cells.

Fractal analysis can also be used to characterize the complexity of gray-scale variations associated with texture. Zheng and Chan [34] used fractals in a preprocessing step to select abnormal regions in mammograms. Guo et al. [63] computed FD to characterize the complexity of ROIs in mammograms, and used a support vector machine for the detection of abnormal regions related to breast masses. Caldwell et al. [64] and Byng et al. [65] computed the FD of breast tumors by applying a modified box-counting method that represents gray-scale values of the surfaces of the tumors as boxes of variable height (see Section 4.3.1). Such a fractal measure can be used to represent the complexity of density variations and texture in breast tissue. Byng et al. [65] showed that a gray-scale-based fractal measure may be used to complement histogram skewness in order to relate breast density to the risk of development of breast cancer. Rangayyan et al. [66, 67] and Banik et al. [68] used FD to distinguish between regions in prior mammograms containing architectural distortion and FP regions detected via phase-portrait analysis.

Li et al. [69] evaluated four approaches to estimate measures of FD, including a conventional box-counting method modified box-counting technique using linear discriminant analysis (LDA), global Minkowski dimension, and modified Minkowski technique using LDA. The FD-based texture features were extracted from ROIs to assess mammographic parenchymal patterns. It was shown that the features could yield radiographic markers to assess the risk of development of breast cancer. Li et al. [70] also evaluated the usefulness of power spectral analysis (PSA) of mammographic parenchymal patterns in the assessment of the risk of development of breast cancer, using manually selected ROIs. The parameter obtained via PSA demonstrated a statistically significant difference between the high-risk and low-risk groups of women in the study.

Other works have reported on the use of FD as a feature for the classification of tumors in magnetic resonance images of the brain [71], ultrasonic images of the liver [72], and images related to colonic cancer [73]. Lee et al. [74] compared several shape factors, including FD, in a study of the irregularity of the borders of melanocytic lesions. Kikuchi et al. [75] investigated the change in FD at different stages of ovarian tumor growth. Nam and Choi [76] computed the FD of regions in mammograms using the box-counting method and found that regions with higher FD indicated the presence of calcification. Klonowski et al. [77] applied Higuchi's method for fractal analysis of the texture seen in histological images as well as for the analysis of the shape of breast masses. Tambasco et al. [78] applied fractal analysis for quantitative analysis of the architectural complexity of microscopic images of histology specimens related to prostate cancer and breast cancer; they found that FD demonstrated statistically highly significant differences between specimens of normal and cancerous tissue.

4.2   FRACTAL ANALYSIS OF SHAPE

One of the aims of the present work is to employ fractal analysis for the classification of breast masses by using their contours. Even though fractal analysis has been widely used in the analysis of biomedical images, only a few studies have specifically applied the method to study and classify mammographic masses [18]. FD may be used as a quantitative measure of the complexity of the contour or boundary of an object. Benign masses and malignant tumors differ significantly in shape complexity, and therefore, it should be possible to differentiate between them by using FD. In the present work, estimates of FD were obtained from 1D signatures of contours of breast masses as well as the contours in their usual 2D forms, by using the ruler method and the box-counting method. These methods to compute the FD, as well as the PSA method, are described in the following subsections.

4.2.1   THE RULER METHOD

One of the popular methods for calculating FD is the ruler method (also known as the compass or divider method) [42]. With different lengths of rulers, the total length of a contour or pattern can be estimated to different levels of accuracy. When using a large ruler, the small details in a given contour would be skipped, whereas when using a small ruler, the finer details would get measured. The estimate of the length improves as the size of the ruler decreases. The compass dimension or FD is obtained from the linear slope of a plot of the log of the measured length vs. the log of the measuring unit.

The formulation of the ruler method is as follows. Let u be the length measured with the compass setting or ruler size s. The value $1/s$ is used to represent the precision of measurement. The power law expected to be satisfied in this case is

$$u = c \, \frac{1}{s^d} \, , \tag{4.4}$$

where c is a constant of proportionality, and the power d is related to FD, expressed here as D, as [42]

$$D = 1 + d \, . \tag{4.5}$$

Applying the log transformation to Equation 4.4 gives

$$\log(u) = \log(c) + d \, \log(1/s) \, . \tag{4.6}$$

Thus, the absolute value of the slope (of the straight-line approximation) of a plot of $\log(u)$ vs. $\log(s)$ can provide an estimate of FD as $D = 1 + d$.

If one denotes $u = ns$, where n is the number of times the ruler is used to measure the length u with the ruler of size s, then

$$\log(n) = \log(c) + (1 + d) \, \log(1/s) \, . \tag{4.7}$$

The absolute value of the slope (of the straight-line approximation) of a plot of $\log(n)$ vs. $\log(s)$ provides an estimate of D directly.

An illustration of the application of the ruler method to compute the FD of the 1D signature of a benign breast mass is shown in Figure 4.7. The related log–log curve is shown in Figure 4.8, along with the straight-line fit derived to estimate FD. Figure 4.9 shows application of the ruler method to compute the FD of the 2D contour of a malignant breast tumor. Examples of breast mass regions, their contours, and the corresponding values of FD are shown in Figure 2.3. It is seen, as expected, that FD increases in value with increasing roughness of the contours.

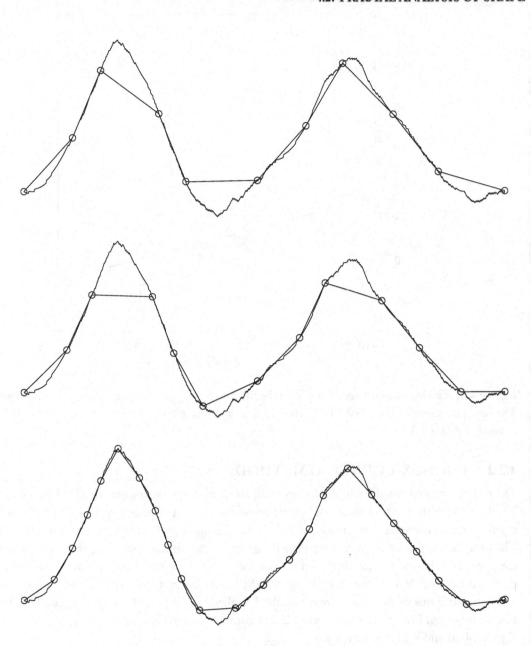

Figure 4.7: Illustration of the ruler method applied to the normalized 1D signature of a benign breast mass to compute its FD. The estimated FD is 1.15.

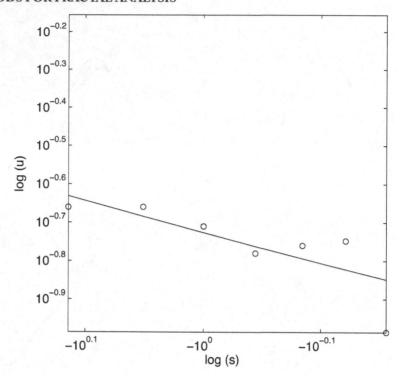

Figure 4.8: The log–log curve resulting from the ruler method applied to the 1D signature in Figure 4.7. The absolute value of the slope of the straight-line approximation of this log–log curve is 0.15; the estimated FD is 1.15.

4.2.2   THE BOX-COUNTING METHOD

The most commonly used method for estimating FD is the box-counting method [42, 79, 80, 81, 82]. The box-counting method consists of partitioning the pattern or image space into square boxes of equal size, and counting the number of boxes that contain a part (at least one pixel) of the image. The process is repeated by partitioning the image space into smaller and smaller squares. The log of the number of boxes counted is plotted against the log of the magnification index for each stage of partitioning. The slope of the best-fitting straight line to the plot gives the FD of the pattern.

Illustrations of the box-counting method applied to compute the FD of a sample 1D signature of a benign breast mass and a sample 2D contour of a malignant breast tumor are shown in Figures 4.10 and 4.11, respectively.

4.2.3   THE POWER SPECTRAL ANALYSIS METHOD

The PSA method is another commonly used method to compute the FD of signals via their frequency content [42, 48, 50, 60, 66, 67, 68]. A random function in time, $V(t)$, may be characterized by its

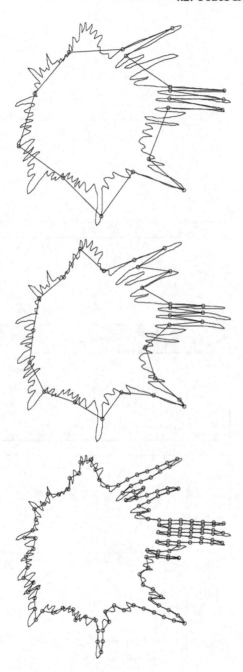

Figure 4.9: Illustration of the ruler method applied to the normalized 2D contour of a malignant breast tumor to compute its FD. The estimated FD is 1.43.

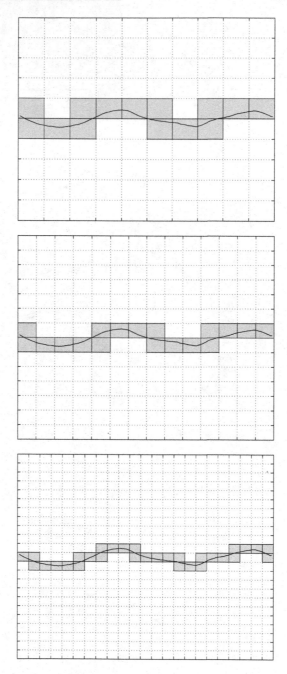

Figure 4.10: Illustration of the box-counting method applied to the normalized 1D signature of a benign breast mass to compute its FD. The estimated FD is 1.04.

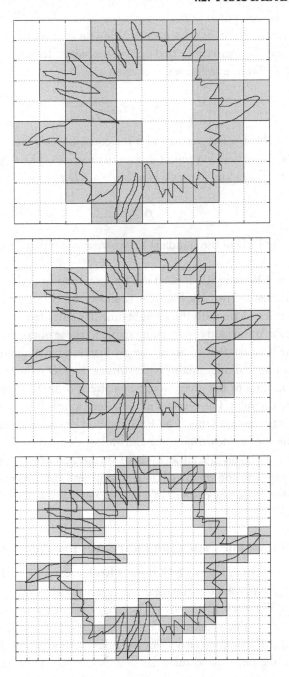

Figure 4.11: Illustration of the box-counting method applied to the normalized 2D contour of a malignant breast tumor to compute its FD. The estimated FD is 1.27.

PSD. The PSD can be computed as the squared magnitude of the Fourier transform of the signal, $V(t)$, or as the Fourier transform of the autocorrelation function of the signal. The PSD characterizes the strength of the frequency variations that exist in the signal. In general, the PSD of a noise process is inversely proportional to the frequency, expressed as

$$S_V(f) \propto \frac{1}{f^\beta}, \tag{4.8}$$

where $S_V(f)$ is the PSD, f is frequency, and β is a constant scale factor. This relationship is known as the $1/f$ noise model and is widely found in natural processes. The slope of the best-fitting straight line to the PSD plotted on a log–log scale is related to FD as [48, 49]

$$D = \frac{5 - \beta}{2}. \tag{4.9}$$

For example, the PSD of the standard Brownian motion pattern (having $H = 1/2$), with FD of 1.5, varies as $1/f^2$.

Rangayyan and Oloumi [50] applied the PSA method to derive the FD of 1D signatures of contours of breast masses and tumors. Figure 4.12 shows the 1D signature of the contour of a benign breast mass and its PSD, along with the straight-line fit computed to estimate FD via PSA. It was shown that FD via PSA provided results comparable to those obtained using the box-counting and ruler methods in the classification of benign breast masses vs. malignant tumors.

The box-counting and ruler methods to estimate the FD require multiscale analysis, achieved by iterative downsampling of the given image or pattern or by iteratively increasing or decreasing the size of the box or the ruler. The number of such iterations is limited by the resolution of the data and the number of samples available. The PSA method provides an alternative and advantageous approach without the need for downsampling and multiscale analysis. The signal processing approach of the PSA method facilitates fractal analysis of images, contours, and patterns with a different perspective as compared to the popular box-counting and ruler methods. Extension of the PSA method to analyze the frequency content of a gray-scale image and to compute its FD is described in Section 4.3.2.

4.2.4   EVALUATION OF THE METHODS WITH TEST PATTERNS

Two simulated test patterns of known FD, the 2D Koch curve [83] and the 1D test pattern defined by Dubuc et al. [79], were used to validate the programs developed to estimate FD. The two test patterns are shown in Figures 4.13 and 4.14. The range of the box size that yielded the most accurate estimate of the FD of the Koch curve was determined to be [1/4, 1/8, 1/16, 1/32, 1/64, 1/128] as a fraction of the size of the original curve. For the ruler method, using the 1D test pattern, the most appropriate range of the normalized ruler size was determined to be [0.05, 0.075, 0.1, . . . , 0.2], as a fraction of the size of the original pattern. Note that, although the test curves are ideally expected to demonstrate fractal characteristics at all scales, practical limitations in their representation and analysis using finite datasets lead to limitations as above.

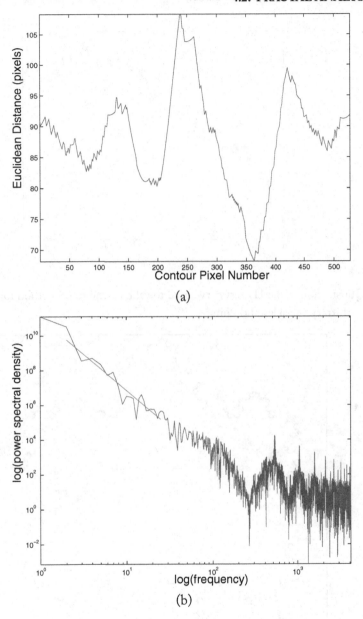

(a)

(b)

Figure 4.12: (a) The 1D signature of the 2D contour of a benign breast mass. The pixel size is 50 μm. (b) PSD of the signature and the straight-line fit computed to estimate FD via PSA. Reproduced with permission from R. M. Rangayyan and F. Oloumi, "Fractal analysis and classification of breast masses using the power spectra of signatures of contours," *Journal of Electronic Imaging*, 21(2):023018, 2012. © SPIE.

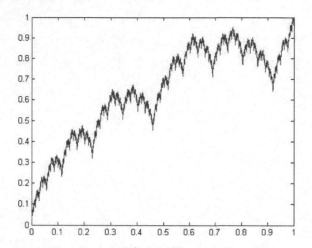

Figure 4.13: Illustration of the 1D test curve used to validate the ruler method for computing FD. The axes represent arbitrary variables and units.

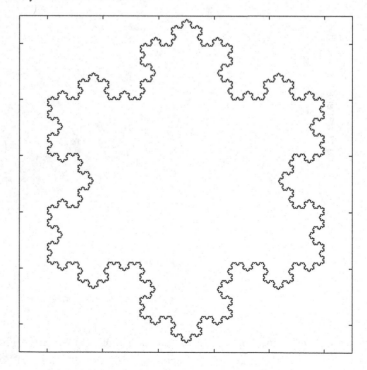

Figure 4.14: Illustration of the 2D Koch snowflake used to validate the box-counting method for computing FD.

It is important to note that, due to practical limitations, it is necessary to limit the range of the reduction factor or measurement scale to a viable range [42, 81]. This caution is applicable to all of the methods described in this chapter for computing the FD. Coelho et al. [81] reported on a study on the need to determine carefully the approximate linear region of the log–log curve from which the slope is to be computed to obtain the correct value of FD. The log–log curve was observed to exhibit two distinct regions, one where FD was incorrectly represented because the magnification factor was too small, and the other a linear region where FD was correctly represented. The ruler size in the ruler method is analogous to the box partitioning size in the box-counting method; it is important to determine a suitable range of the ruler size to estimate FD accurately [42]. This step also accommodates limitations in the fractal characteristics of a given pattern due to image size and sampling considerations. Rangayyan and Oloumi [50] described the need to limit the range of frequencies over which the straight-line fit should be computed to the PSD in the estimation of FD using the PSA method.

Rangayyan and Oloumi [50] evaluated estimates of the FD of a number of fBm signals using the box-counting, ruler, and PSA methods. The results indicated that the PSA method is the best suited method for the estimation of the FD of fBm and self-affine signals, among the methods studied.

4.3   FRACTAL ANALYSIS OF GRAY-LEVEL IMAGES

Fractal analysis can be used to classify breast masses as benign or malignant based on the gray-scale variability (texture and roughness) exhibited in their images, with the former being mostly homogeneous and the latter showing heterogeneous texture. This section provides descriptions of two methods that can be used to compute the FD of breast mass regions.

4.3.1   THE BLANKET METHOD

The blanket method, presented by Caldwell et al. [64] and Byng et al. [65], is analogous to the box-counting method. In this method, a gray-scale image is treated as having a third dimension, that is, the gray-scale value of each pixel is treated as having a vertical height. The surface area of the image is estimated using a range of areas or pixel sizes. Different pixel sizes are synthesized by averaging adjacent pixels together. Depending on the pixel size, ε, that is used, the surface area, $A(\varepsilon)$, yields different estimates. The surface area $A(\varepsilon)$ is computed as the sum of the area of each pixel of size ε, plus the sum of the exposed area between each pixel and its neighboring pixels due to their differences in heights. This process is illustrated in Figure 4.15. For an image of size $N \times N$ pixels, with each pixel having the same value and size $\varepsilon \times \varepsilon$ units of area, the surface area is equal to $(N\varepsilon)^2$. When adjacent pixels possess different values, more surface area of the blocks representing the pixels is exposed, as shown in Figure 4.15. The total surface area for the image is given by

$$A(\varepsilon) = \sum_{m=0}^{N-2} \sum_{n=0}^{N-2} \{\varepsilon^2$$
$$+ \;\; \varepsilon\,[\,|f_\varepsilon(m,n) - f_\varepsilon(m,n+1)| + |f_\varepsilon(m,n) - f_\varepsilon(m+1,n)|\,]\,\}, \qquad (4.10)$$

where $f_\varepsilon(m,n)$ is the 2D image expressed as a function of the pixel size ε. A power-law relationship exists between $A(\varepsilon)$ and ε for images exhibiting fractal characteristics [64, 65]. The value of FD is given by

$$D = 2 - \frac{\Delta \log[A(\varepsilon)]}{\Delta \log[\varepsilon]}. \qquad (4.11)$$

The fractal model is appropriate if, over a range of pixel sizes, the regression between $\log[A(\varepsilon)]$ and $\log[\varepsilon]$ is linear.

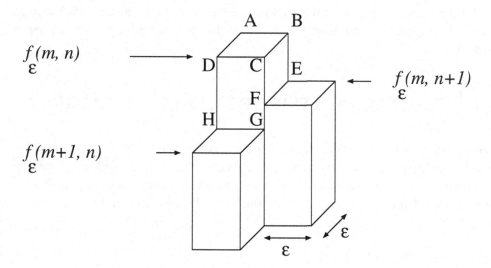

Figure 4.15: Illustration of the computation of the surface area of a gray-scale image as part of the blanket method to compute the FD of an image. The exposed area of the pixel at (m,n) is the net area of the parts labeled as ABCD, BEFC, and DCGH. Reproduced with permission from R. M. Rangayyan, *Biomedical Image Analysis*, CRC Press, Boca Raton, FL, 2005. © CRC Press.

4.3.2   POWER SPECTRAL ANALYSIS OF GRAY-LEVEL IMAGES

The PSA method can be used to estimate the FD of images [50, 66, 67, 68, 84, 85, 86]. The principle applied in this method is similar to the method described in Section 4.2.3. The steps to compute the FD for a gray-scale image using PSA are as follows.

1. Compute the Fourier transform of the image; zero pad the image prior to computation of the Fourier transform, if necessary.

2. Compute the 2D PSD by taking the squared magnitude of the Fourier transform obtained in the previous step.

3. Transform the 2D PSD into a 1D function by radial averaging.

4. Fit a straight line to a selected range of frequencies of the 1D function plotted on a log–log scale.

5. Determine the slope, β, of the best-fitting line and compute the FD of the image as

$$FD = \frac{8 - \beta}{2}. \tag{4.12}$$

The PSA method applied to images is different from the PSA method applied to 1D signals only in the step to convert the PSD from 2D to 1D and in the constant term added in the formula to estimate FD. Otherwise, the steps in computing FD using the PSA method are the same when applied to 1D signals or 2D images. Banik et al. [68] described a method to transform the 2D PSD of an image from rectangular (Cartesian) to polar coordinates and subsequent reduction to a 1D function of radial frequency; they also discuss the importance of limiting the range of frequencies used in fitting a linear model for PSA. The work of Banik et al. demonstrates the application of PSA for fractal analysis of mammographic images for the detection of architectural distortion.

4.3.3   EVALUATION OF THE METHODS WITH TEST IMAGES

Images of known FD can be synthesized based on the fBm model. A popular method to generate fractal images is the 2D random midpoint displacement method [42, 48, 49]. This method begins with a square image space bounded by four corner pixels. An edge midpoint, defined as the point half way between the two corresponding corner pixels, is computed by averaging the two corner values plus a random value. The diagonal midpoint, defined as the point where the two diagonals meet, is computed by averaging all four corner values plus a random value. This process is repeated for each smaller square bounded by four pixels.

The blanket method and the PSA method were applied to fractal images of known FD (generated using a program developed by Dr. M. Tambasco, University of Calgary), to verify their accuracy. Three sample test images used for this purpose, with their actual (model) FD and computed FD using the blanket and PSA methods, are shown in Figure 4.16. The estimated FD values have substantial errors in two of the three cases illustrated. Regardless, both the blanket and PSA methods provided higher estimates of FD for images with higher degrees of gray-level variation.

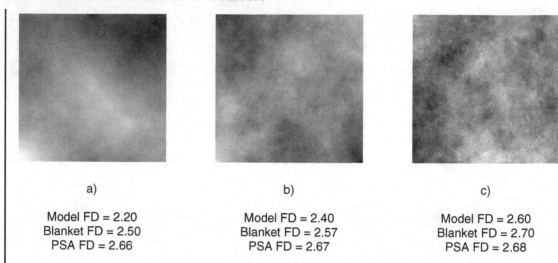

a)

Model FD = 2.20
Blanket FD = 2.50
PSA FD = 2.66

b)

Model FD = 2.40
Blanket FD = 2.57
PSA FD = 2.67

c)

Model FD = 2.60
Blanket FD = 2.70
PSA FD = 2.68

Figure 4.16: Three test images generated based on the fBm model with $H = 0.8, 0.6$, and 0.4, which correspond to $FD = 2.2, 2.4$, and 2.6. The FDs of the images computed using the blanket method and the PSA method are also shown. The images were generated using a program developed by Dr. M. Tambasco.

4.4   REMARKS

This chapter has demonstrated that fractals have several different properties, including self-similarity at all scales, being defined by a recursive process, appearing highly irregular and too complex to be characterized by traditional geometry, and having a nonintegral dimension. This chapter has also presented a description of the fBm model for the purpose of characterizing the randomness found in natural fractal patterns such as clouds and landscapes. Derived from the understanding of fractal geometry are methods for quantitative analysis of the fractal nature of objects via a measure called the FD. This chapter has presented some of the well-known methods for computing the FD, which are the ruler method, box-counting method, blanket method, and PSA method. The ruler method and the box-counting method are applied in the present work to compute the FD of 1D signature representations and 2D contours of breast masses. The blanket method is applied to compute the FD of the gray-scale ROIs and ribbons of breast masses. The present work demonstrates that the various concepts of fractals may be used to characterize not only the complexity of the shapes of contours of breast masses and tumors, but also their complexity associated with gray-scale texture.

CHAPTER 5

Pattern Classification

After measures of the shape and textural characteristics have been derived from contours and regions of breast masses, they may be used in pattern classification experiments to discriminate the breast masses into categories of benign and malignant diseases. This chapter presents descriptions of the pattern recognition techniques and statistical techniques used in the classification experiments presented in this book.

5.1 FISHER LINEAR DISCRIMINANT ANALYSIS

Dimensionality reduction is one of the central issues in statistical data modeling. FLDA is a dimensionality reduction technique that projects multidimensional data onto a 1D space or line which allows for easier discrimination [87]. In a two-class problem, the projection is designed to maximize the distance between the means of the two classes and minimize the variance within each class. FLDA is used in classification problems where feature sets are expected to be linearly separable.

An illustration of a linear classifier based on FLDA is depicted in Figure 5.1. Consider two classes, C_1 and C_2, in which each sample is described in terms of the features x_1 and x_2. Suppose that the feature x_1 is used for classification and all of the data points are projected onto the x_1 axis. This projection creates overlapping regions that contain samples from both classes. This indicates that x_1 provides poor discrimination between the classes C_1 and C_2. Classification using the feature x_2 also yields poor results. However, by visual inspection, it can be seen that the classes C_1 and C_2 are perfectly separable by a slanted straight line. Therefore, a classifier can be built based on projections of the samples in the classes onto the line labeled as F in the illustration in Figure 5.1. FLDA is a formalization of this classification example.

The principle used to formulate FLDA is similar to the analysis of variance (ANOVA) in that they both use the ratio of the between-class scatter to the within-class scatter. FLDA computes the weights that define the projection line by maximizing this ratio, whereas ANOVA uses this ratio and the F distribution to test for significant difference between the means of the classes. The between-class and within-class scatter factors are explained by the following example.

Consider an example of m groups, each with n data points. The aim is to determine whether the means of these groups are significantly different from one another. The null hypothesis states that the means of the groups are the same, whereas the alternative hypothesis states that the means are not all the same. The grand mean is defined as the mean of all of the $m \times n$ data points. It follows that, if the null hypothesis is true, the mean is the same for each group and the deviation of any one

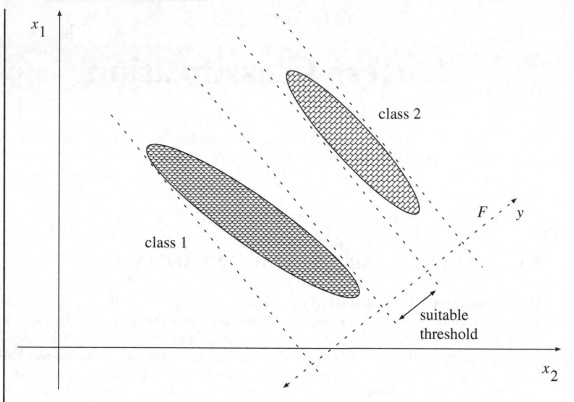

Figure 5.1: Illustration of a classifier based upon the principles of FLDA.

element or data point from the grand mean arises only by chance. On the other hand, if the means are different, there are two reasons why a given data point may deviate from the grand mean.

- The mean of its own group is different from the grand mean.

- There is a chance variation within its own group.

The first reason represents the deviation of each group mean from the grand mean, which is referred to as the between-class scatter factor. The second reason represents the deviation of each data point from its own group mean, which is referred to as the within-class scatter factor.

    FLDA is derived using the following binary classification example [87]. Suppose that a given dataset

$$X = \{\mathbf{x}_1, \mathbf{x}_2, \ldots, \mathbf{x}_N\} = \{X_1, X_2\} \tag{5.1}$$

is partitioned into N_1 training vectors in subset X_1, corresponding to class C_1, and N_2 training vectors in subset X_2, corresponding to class C_2, where $N_1 + N_2 = N$. The projection of $\mathbf{x}_i$ onto the FLDA discriminant line is given as

$$y_i = \mathbf{w}^T \mathbf{x}_i, \tag{5.2}$$

where the corresponding projection y_i belongs to the subset Y_1 or Y_2, and $\mathbf{w}$ is the weight vector that defines the direction of the projected values y_i. The mean of each class prior to projection is

$$\mathbf{m}_k = \frac{1}{N_k} \sum_{\mathbf{x}_i \in X_k} \mathbf{x}_i, \tag{5.3}$$

where $k = 1, 2$, and $i = 1, 2, \ldots, N$. The mean for each class after projection is

$$
\begin{aligned}
\tilde{m}_k &= \frac{1}{N_k} \sum_{y_i \in Y_k} y_i \\
&= \frac{1}{N_k} \sum_{\mathbf{x}_i \in X_k} \mathbf{w}^T \mathbf{x}_i, \\
&= \mathbf{w}^T \mathbf{m}_k, \quad k = 1, 2.
\end{aligned}
\tag{5.4}
$$

The difference between the projected means of the classes is

$$|\tilde{m}_1 - \tilde{m}_2| = |\mathbf{w}^T \mathbf{m}_1 - \mathbf{w}^T \mathbf{m}_2|. \tag{5.5}$$

The scatter of the projected sample y_i with respect to each class is

$$
\begin{aligned}
\tilde{s}_k^2 &= \sum_{y_i \in Y_k} (y_i - \tilde{m}_k)^2 \\
&= \sum_{\mathbf{x}_i \in X_k} (\mathbf{w}^T \mathbf{x}_i - \mathbf{w}^T \mathbf{m}_k)^2 \\
&= \sum_{\mathbf{x}_i \in X_k} \mathbf{w}^T (\mathbf{x}_i - \mathbf{m}_k)(\mathbf{x}_i - \mathbf{m}_k)^T \mathbf{w}, \quad k = 1, 2 \\
&= \mathbf{w}^T \mathbf{S}_k \mathbf{w}, \tag{5.6}
\end{aligned}
$$

where the scatter matrix $\mathbf{S}_k$ is defined as

$$\mathbf{S}_k = \sum_{\mathbf{x}_i \in X_k} (\mathbf{x}_i - \mathbf{m}_k)(\mathbf{x}_i - \mathbf{m}_k)^T, \quad k = 1, 2, \tag{5.7}$$

and the matrix $\mathbf{S}_W$ is defined as

$$\mathbf{S}_W = \mathbf{S}_1 + \mathbf{S}_2. \tag{5.8}$$

The sum of the scatter for the two classes can be written as

$$\tilde{s}_1^2 + \tilde{s}_2^2 = \mathbf{w}^T \mathbf{S}_W \mathbf{w}. \tag{5.9}$$

The FLDA criterion function

$$J(\mathbf{w}) = \frac{|\tilde{m}_1 - \tilde{m}_2|^2}{\tilde{s}_1^2 + \tilde{s}_2^2} \tag{5.10}$$

leads to the best separation between the classes when it is at its maximum. The derivation for the denominator term of this function has been shown above. The numerator term of this function describes the separation of the projected means of the two classes, and is given as

$$
\begin{aligned}
(\tilde{m}_1 - \tilde{m}_2)^2 &= (\mathbf{w}^T \mathbf{m}_1 - \mathbf{w}^T \mathbf{m}_2)^2 \\
&= \mathbf{w}^T (\mathbf{m}_1 - \mathbf{m}_2)(\mathbf{m}_1 - \mathbf{m}_2)^T \mathbf{w} \\
&= \mathbf{w}^T \mathbf{S}_B \mathbf{w} .
\end{aligned} \tag{5.11}
$$

Therefore, the criterion function can be expressed in terms of the between-class scatter, $\mathbf{S}_B$, and the within-class scatter, $\mathbf{S}_W$, written as

$$J(\mathbf{w}) = \frac{\mathbf{w}^T \mathbf{S}_B \mathbf{w}}{\mathbf{w}^T \mathbf{S}_W \mathbf{w}} . \tag{5.12}$$

This expression resembles the generalized Rayleigh quotient. The weight vector $\mathbf{w}_o$ that maximizes this expression can be calculated by solving a generalized eigenvalue problem. When a new sample is to be classified, its feature vector $\mathbf{x}$ is projected using Equation 5.2 and two-category classification is performed as

$$\mathbf{x} \in \begin{cases} C_1 & \text{if } y = \mathbf{w}_o^T \mathbf{x} < T_1 \\ C_2 & \text{otherwise} \end{cases} , \tag{5.13}$$

where T_1 is a threshold.

5.2   THE BAYESIAN CLASSIFIER

Bayesian analysis is a statistical approach to classification which identifies the optimal decision as one that carries the minimal overall probability of risk [87]. Bayes' theorem is derived from the principles of conditional probability. Let ω_1 and ω_2 be two mutually exclusive events in the sample space S, and let x be any event of the sample space such that $p(x) \neq 0$. The probability of event x occurring given that event ω_1 has already occurred is referred to as the conditional probability of x given ω_1, denoted as $p(x|\omega_1)$, and given as

$$p(x|\omega_1) = \frac{p(x, \omega_1)}{P(\omega_1)}, \tag{5.14}$$

where $p(x, \omega_1)$ is the joint probability of x and ω_1. Because $p(x, \omega_1) = p(\omega_1, x)$,

$$p(x|\omega_1)P(\omega_1) = P(\omega_1|x)p(x). \tag{5.15}$$

Rearranging Equation 5.15 leads to Bayes' formula:

$$P(\omega_1|x) = \frac{p(x|\omega_1)P(\omega_1)}{p(x)}, \tag{5.16}$$

where, by the theorem of mutually exclusive events,

$$p(x) = \sum_{j=1}^{2} p(x|\omega_j)P(\omega_j). \tag{5.17}$$

Therefore, a generalized Bayes' formula is expressed as

$$P(\omega_j|x) = \frac{p(x|\omega_j)P(\omega_j)}{\sum_{j=1}^{c} p(x|\omega_j)P(\omega_j)}, \tag{5.18}$$

where c is the total number of events (states of nature or classes). The posterior probability, $P(\omega_j|x)$, is the probability that the true state is ω_j given that x was observed. The likelihood of ω_j with respect to x, expressed as $p(x|\omega_j)$, indicates that the category ω_j for which $p(x|\omega_j)$ is large is more likely to be the true category. The prior probability, $P(\omega_j)$, is the prior knowledge of the probability of the events. The evidence factor, $p(x)$, is the probability of the evidence x irrespective of any knowledge about the state of nature ω_j. Bayes' formula can be also be expressed as:

$$posterior = \frac{likelihood \times prior}{evidence}. \tag{5.19}$$

A loss function represents the cost associated with a certain action. Let $\{\alpha_1, \alpha_2, \ldots, \alpha_a\}$ be a set of a possible actions and $\{\omega_1, \omega_2, \ldots, \omega_c\}$ be a set of c states of nature. The loss function is expressed as $\lambda(\alpha_i | \omega_j)$ and describes the loss incurred for taking action α_i, which corresponds to deciding on ω_i, when the true state of nature is ω_j. Suppose that a particular set of events $\mathbf{x}$ was observed, the action α_i was planned to be taken, and $P(\omega_j | \mathbf{x})$ is the probability that the true state of nature is ω_j; then, the expected loss or conditional risk is defined as

$$R(\alpha_i | \mathbf{x}) = \sum_{j=1}^{c} \lambda(\alpha_i | \omega_j) P(\omega_j | \mathbf{x}) . \tag{5.20}$$

Bayes' decision rule states that, to minimize the overall risk, one should compute the conditional risk $R(\alpha_i | \mathbf{x})$ for all of the possible actions, $i = 1, 2, \ldots, a$, and then select the action α_i that is associated with the lowest conditional risk.

In a binary classification problem in which there are two classes, ω_1 and ω_2, and two corresponding actions, α_1 and α_2, using the Bayesian minimum-risk decision rule, one decides on ω_1 if

$$R(\alpha_1 | \mathbf{x}) < R(\alpha_2 | \mathbf{x}) , \tag{5.21}$$

and on ω_2 otherwise. The conditional risk for choosing either class is expressed as

$$R(\alpha_1 | \mathbf{x}) = \lambda(\alpha_1 | \omega_1) P(\omega_1 | \mathbf{x}) + \lambda(\alpha_1 | \omega_2) P(\omega_2 | \mathbf{x}) , \tag{5.22}$$

$$R(\alpha_2 | \mathbf{x}) = \lambda(\alpha_2 | \omega_1) P(\omega_1 | \mathbf{x}) + \lambda(\alpha_2 | \omega_2) P(\omega_2 | \mathbf{x}) . \tag{5.23}$$

Using Bayes' formula, Equation 5.21 is expressed in terms of the likelihoods and prior probabilities instead of the posterior probabilities as

$$[\lambda(\alpha_2 | \omega_1) - \lambda(\alpha_1 | \omega_1)] p(\mathbf{x} | \omega_1) P(\omega_1) > [\lambda(\alpha_1 | \omega_2) - \lambda(\alpha_2 | \omega_2)] p(\mathbf{x} | \omega_2) P(\omega_2) . \tag{5.24}$$

Rearranging the above equation, the condition to select ω_1 becomes

$$\frac{p(\mathbf{x} | \omega_1)}{p(\mathbf{x} | \omega_2)} > \frac{[\lambda(\alpha_1 | \omega_2) - \lambda(\alpha_2 | \omega_2)]}{[\lambda(\alpha_2 | \omega_1) - \lambda(\alpha_1 | \omega_1)]} \frac{P(\omega_2)}{P(\omega_1)} . \tag{5.25}$$

It is seen that a binary classifier based on the Bayesian minimum-risk decision rule can be interpreted as calling for deciding on class ω_1 if the likelihood ratio $p(\mathbf{x} | \omega_1) / p(\mathbf{x} | \omega_2)$ exceeds a threshold value.

5.3 RECEIVER OPERATING CHARACTERISTICS

ROC analysis [88] is a useful technique to evaluate binary classifiers by graphical visualization. It is most commonly used to evaluate the performance of medical tests in discriminating diseased cases (positive) from nondiseased cases (negative).

In a binary classification system, the cases are classified into one of either two categories or classes based on a decision threshold. In an ideal situation, the two classes are perfectly separable, and a certain range of decision thresholds or cut-off values provide perfect discrimination between the classes. However, in most instances, the distributions of the two classes overlap, and every cut-off value selected results in some erroneous predictions, with diseased cases misclassified as nondiseased and nondiseased cases misclassified as diseased. From the schematic case shown in Figure 5.2, the following are four possible outcomes:

- True positive (TP)—the number of positive cases correctly classified as positive.

- True negative (TN)—the number of negative cases correctly classified as negative.

- False positive (FP)—the number of negative cases incorrectly classified as positive.

- False negative (FN)—the number of positive cases incorrectly classified as negative.

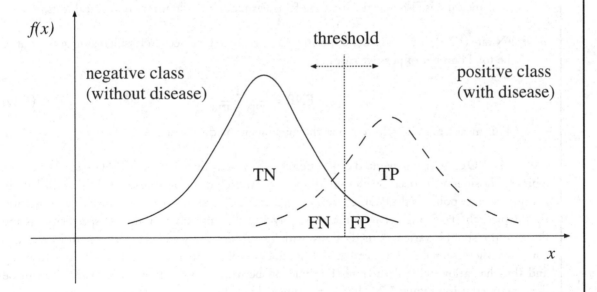

Figure 5.2: Illustration of the distribution of two classes and the four possible outcomes as the threshold is varied.

Based on the outcomes listed above, the following statistics are defined.

- TP rate (TPR) or sensitivity—the number of positive cases that are correctly classified as positive out of the total number of positive cases.

$$TPR = \frac{TP}{TP + FN}.$$ (5.26)

This measure is also referred to as the hit rate or recall rate in medical tests.

- TN rate (TNR) or specificity—the number of negative cases that are correctly classified as negative out of the total number of negative cases.

$$TNR = \frac{TN}{TN + FP}.$$ (5.27)

- FP rate (FPR)—the number of negative cases that are incorrectly classified as positive out of the total number of negative cases.

$$FPR = \frac{FP}{TN + FP}.$$ (5.28)

This measure is also referred to as the false alarm rate or fallout rate in medical tests.

- FN rate (FNR)—the number of positive cases that are incorrectly classified as negative out of the total number of positive cases.

$$FNR = \frac{FN}{TP + FN}.$$ (5.29)

This measure is also referred to as the miss rate in medical tests.

The ROC curve is generated by plotting the TPR (sensitivity) vs. the FPR ($1-$ specificity) for a binary classifier as its decision threshold is varied. At each decision threshold, TPR and FPR are computed, and a point on the ROC curve is formed. ROC analysis provides a graphical representation that depicts the tradeoffs between the sensitivity and the false-alarm rate ($1-$ specificity) as the decision threshold is varied. A medical test with a high sensitivity will classify more diseased cases correctly, whereas a medical test with a high specificity will classify more nondiseased cases correctly and thus has a low false-alarm rate. The tradeoff between the sensitivity and specificity can be demonstrated using Figure 5.2. Moving the threshold to the left increases the number of TP (thus reduces the number of FN) but decreases the number of TN (thus increases the number of FP). This results in a higher sensitivity (due to more TP), but a lower specificity (due to decreased TN). The reverse occurs when the threshold is moved to the right: the number of TP decreases, resulting in lower sensitivity, while the number of TN increases, resulting in a higher specificity. A good classifier will have high sensitivity as well as high specificity.

The performance of a classifier can be measured by the AUC, which ranges from 0 to 1. An area of 1 indicates a perfect separation between the positive and the negative classes. An area of 0.5 indicates a complete overlap between the distributions of the two classes. An area less than 0.5 does not necessarily indicate a poor classifier. By reversing the definition of the classes, that is, the positive classes are labeled as negative and the negative classes are labeled as positive, the ROC curve is reflected about its diagonal and the AUC becomes greater than 0.5. The AUC is a useful measure of a classifier's performance that can be easily interpreted.

5.4   THE t-TEST AND p-VALUE

The t-test is used to assess whether the means of two groups are statistically different from each other [89, 90]. It does so by assessing the difference between the means of the two groups relative to the spread or variability of their values. Assuming that the variance, σ^2, is the same for the two groups, the t-statistic, t_s, is computed as

$$t_s = \frac{\bar{x}_1 - \bar{x}_2}{sp \times \sqrt{\frac{1}{n_1} + \frac{1}{n_2}}} , \tag{5.30}$$

where $\bar{x}_1$ and n_1 are the mean and size of the first group, $\bar{x}_2$ and n_2 are the mean and size of the second group, and sp is the pooled standard deviation, defined as

$$sp = \sqrt{\frac{(n_1 - 1)\sigma^2 + (n_2 - 1)\sigma^2}{n_1 + n_2 - 2}} . \tag{5.31}$$

If the variance is not the same for the two groups, t_s is computed as

$$t_s = \frac{\bar{x}_1 - \bar{x}_2}{\sqrt{\frac{\sigma_1^2}{n_1} + \frac{\sigma_2^2}{n_2}}} , \tag{5.32}$$

where σ_1^2 is the variance of the first group and σ_2^2 is the variance of the second group.

The number of degrees of freedom is equal to $n_1 + n_2 - 2$. The significance level is often set at 0.05; this means that, 5 times out of a 100, the difference between the means is concluded to be statistically significant when it is not actually so. It is the probability of being incorrect if the null hypothesis (which assumes that $\bar{x}_1 = \bar{x}_2$) is rejected. Given the degrees of freedom and significance level, the T-critical value, T_c, can be determined using the t-distribution look-up table. When t_s, computed with either Equation 5.32 or Equation 5.30, is larger than T_c, the null hypothesis is rejected.

The t-distribution curve shown in Figure 5.3 illustrates the significance level and p-value in terms of the area under the t-distribution curve. The significance level is the area under the curve to the right of the positive T_c, plus the area to the left of the negative T_c (all of the areas labeled as A and B). The p-value is the area under the curve to the right of the positive t_s, plus the area to the left of the negative t_s (all of the areas labeled as B).

The p-value is a statistical measure of the probability that the results observed in a study could have occurred by chance. A small p-value is a rejection of the null hypothesis in favor of the alternative hypothesis (which assumes that $\bar{x}_1 \neq \bar{x}_2$) because it indicates how unlikely it is that a test statistic as extreme as or more extreme than the one given by the data will be observed from the population if the null hypothesis were true. For example, a p-value of 0.01 indicates that there is a 1 out of a 100 chance that the result occurred by chance. Any test resulting in a p-value under 0.05 is considered to indicate that the difference between the means is statistically significant, and a p-value under 0.01 to indicate that the difference between the means is statistically highly significant.

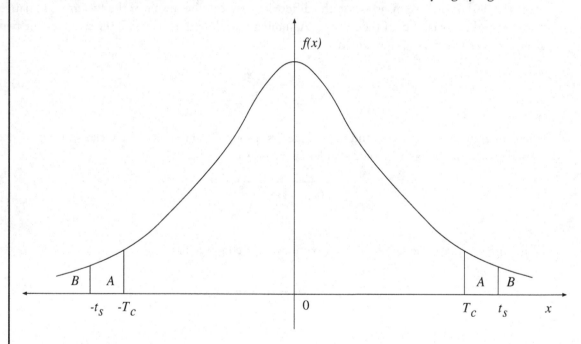

Figure 5.3: The t-distribution curve.

5.5   REMARKS

In this chapter, two well-known pattern classification methods, FLDA and the Bayesian classifier, were described. Such methods may be used to classify or categorize breast masses based on their features. This chapter also presented a brief description of the method of ROC analysis, which is commonly used to evaluate the performance of a medical test by providing a graphical representation of the trade-offs between the sensitivity and the false-alarm rate. Furthermore, the AUC quantifies the overall ability of the test to discriminate between diseased and nondiseased cases. The t-test and p-value were also described in this chapter and are applied in the present study to evaluate whether observed differences bear statistical significance.

CHAPTER 6

Results of Classification of Breast Masses

This chapter presents the results of classification of breast masses based on the methods described in the preceding chapters. For classification experiments using a single feature, the ROC analysis method was used with a sliding threshold applied directly to the feature. For classification experiments using multiple shape or texture features, the FLDA or Bayesian classifier was applied, and the resulting discriminant values were used as inputs for ROC analysis. The AUC was computed to serve as a measure of classification performance.

Section 6.1 shows the results of sorting contours of breast masses in terms of increasing values of FD. Section 6.2 presents the results of shape analysis using FD alone and in conjunction with some of the other shape measures described in Chapter 2. Section 6.3 presents the results of analysis of gray-scale variation using FD and the results of analysis of the effect of pixel size on Haralick's texture measures in the classification of ROIs and ribbons of breast masses.

6.1 RANK-ORDERING CONTOURS OF MASSES BY FRACTAL DIMENSION

FD was computed for the 1D signatures derived from the contours of the breast masses in the first and second datasets. Figure 6.1 shows the contours of the 57 masses in the first dataset, ranked in the order of increasing FD obtained by the ruler method applied to their 1D signatures. Malignant tumors generally have higher FD because they are more jagged and spiculated than benign masses. The average FD of the 37 benign masses in the first dataset is 1.13 ± 0.05 (mean $\pm$ standard deviation). The average FD of the 20 malignant tumors in the first dataset is 1.41 ± 0.15.

Figure 6.2 shows the contours of the 54 masses in the second dataset, ranked in the order of increasing FD obtained by the ruler method applied to their 1D signatures. The average FD of the 28 benign masses in the second dataset is 1.22 ± 0.09. The average FD of the 26 malignant tumors in the second dataset is 1.35 ± 0.11.

Visual inspection of the contours sorted based on increasing values of FD, seen in Figures 6.1 and 6.2, confirms that the contours are arranged in the order of most smooth or circumscribed to most jagged or spiculated, as expected, in most cases. This result is encouraging, indicating that FD can be used to differentiate between benign masses and malignant tumors in mammograms.

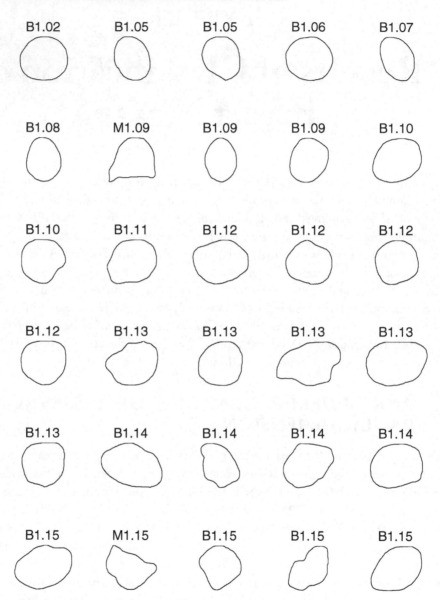

Figure 6.1: Contours of the 37 benign masses and the 20 malignant tumors in the first dataset, ranked by their FD estimated by the ruler method applied to their 1D signatures. The contours are of widely differing size, but have been scaled to the same bounding box in the plots. B = Benign, M = Malignant. Reproduced, with kind permission from Springer Science+Business Media B. V., from R. M. Rangayyan and T. M. Nguyen, "Fractal analysis of contours of breast masses in mammograms," *Journal of Digital Imaging*, 20(3):223–237, 2007. © Springer.   *Continues.*

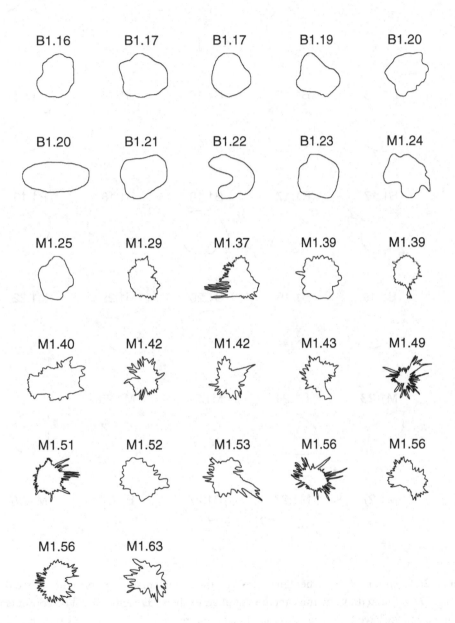

Figure 6.1: *Continued.*

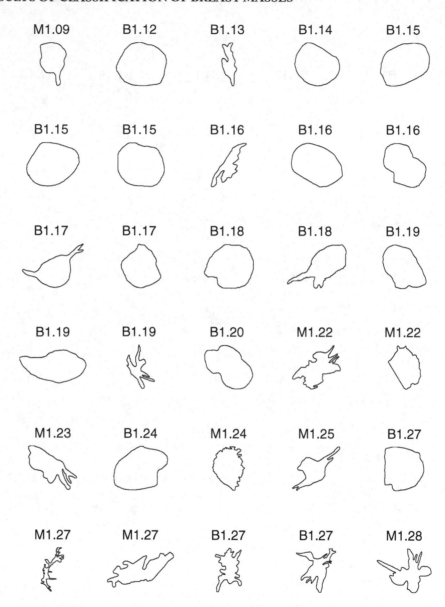

Figure 6.2: Contours of the 28 benign masses and the 26 malignant tumors in the second dataset, ranked by their FD estimated by the ruler method applied to their 1D signatures. The contours are of widely differing size, but have been scaled to the same bounding box in the plots. B= Benign, M = Malignant. Reproduced, with kind permission from Springer Science+Business Media B. V., from R. M. Rangayyan and T. M. Nguyen, "Fractal analysis of contours of breast masses in mammograms," *Journal of Digital Imaging*, 20(3):223–237, 2007. © Springer.   *Continues.*

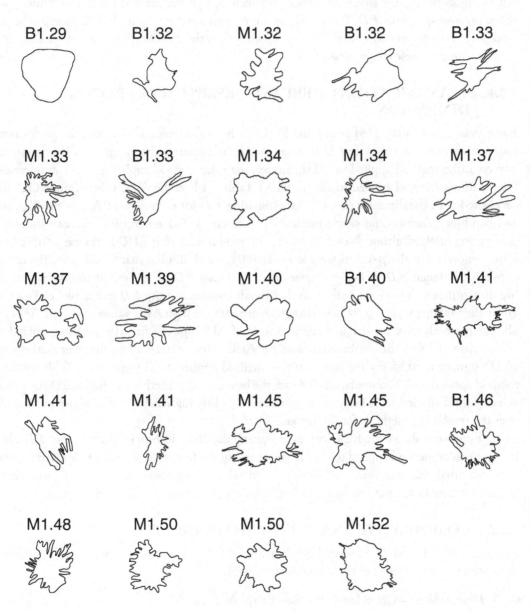

Figure 6.2: *Continued.*

6.2   RESULTS OF SHAPE ANALYSIS

On the basis of notable shape differences between benign masses and malignant tumors, several shape measures, such as FD, C, F_{cc}, SI, and FF, were used for their classification in the present work and previous related works. This section compares the classification performance of each and combinations of these shape measures.

6.2.1   CLASSIFICATION OF BREAST MASSES USING FRACTAL DIMENSION

Rangayyan and Nguyen [18] performed ROC analysis to assess and compare the performance of four methods used to compute FD of breast masses (the ruler method applied to 1D signatures, the box-counting method applied to 1D signatures, the ruler method applied to 2D contours, and the box-counting method applied to 2D contours). Table 6.1 lists the AUC values from ROC analysis conducted with the different methods for estimating FD for each dataset. Across the first, second, and combined datasets, no single method for estimating FD emerged as the consistently best-performing method, although each method gave good results with AUC in the range of 0.75–0.94. Furthermore, all of the p-values were less than 0.01, which implies that for all cases, the difference between the mean FD of benign masses and the mean FD of malignant tumors is statistically highly significant. This is an indication that the distribution of the FD of the two classes (benign and malignant) are separable, which is in agreement with the high AUC values observed. As expected, all four methods gave consistently higher values of AUC for the first dataset as compared to the second dataset. With the combined dataset, the AUC values were: 0.89 for the ruler method applied to 1D signatures, 0.88 for the box-counting method applied to 1D signatures, 0.88 for the ruler method applied to 2D contours, and 0.84 for the box-counting method applied to 2D contours. The ruler method applied to 1D signatures provided a slightly higher AUC, and therefore, was chosen over the remaining methods for further analysis.

Figure 6.3 shows the ROC curves indicating the classification performance of FD obtained using the ruler method applied to the 1D signatures of the first, second, and combined datasets. The FD computed using the ruler method applied to 1D signatures was used in conjunction with other shape measures in the pattern classification studies presented in the following section.

6.2.2   COMPARATIVE ANALYSIS OF SHAPE FACTORS

Table 6.2 lists the AUC of several combinations of the shape measures tested. The results are presented for each dataset in the following paragraphs.

- *First dataset*— Considering the use of a single shape feature, F_{cc} achieved the highest classification accuracy with AUC = 0.99. The other four shape features (C, SI, FF, and FD) also achieved high AUC values in the range of 0.91–0.98. Recall that the first dataset contains mostly typical benign masses with smooth contours and malignant tumors with rough contours, which do not generally cause difficulties in pattern classification studies. The results

Table 6.1: Comparison of the results of the ruler and box-counting methods applied to 1D signatures and 2D contours of breast masses to obtain FD, in terms of AUC. The p-value was less than 0.01 for each feature. Reproduced, with kind permission from Springer Science+Business Media B. V., from R. M. Rangayyan and T. M. Nguyen, "Fractal analysis of contours of breast masses in mammograms," *Journal of Digital Imaging*, 20(3):223–237, 2007. © Springer.

Method	Dataset 1	Dataset 2	Combined
Ruler method applied to 1D signatures	0.91	0.80	0.89
Ruler method applied to 2D contours	0.94	0.81	0.88
Box-counting method applied to 1D signatures	0.89	0.80	0.88
Box-counting method applied to 2D contours	0.90	0.75	0.84

indicate that each of the five shape measures used in the present work can, on its own, accurately differentiate between typical benign masses and malignant tumors. For this dataset, the use of combinations of multiple shape measures did not lead to any higher accuracy than that with F_{cc} alone. This indicates that F_{cc} is an effective shape feature for the classification of typical benign masses and malignant tumors.

• *Second dataset*— The AUC values in all of the experiments conducted for single-feature and multiple-feature classification were lower for the second dataset as compared to those with the first dataset. Recall that the second dataset contains unusually large numbers of atypical masses and tumors. In the case of single-feature classification, FD achieved the highest AUC of 0.80, indicating that it is, on the average, better at classifying the atypical cases such as SB masses and CM tumors than the other shape measures tested. The features SI and F_{cc} provided a lower AUC of 0.77. When FD was combined with the other shape measures, higher values for AUC were obtained as follows: 0.82 with $[FD, F_{cc}]$ or $[FD, SI]$; 0.81 with $[C, FD, F_{cc}]$; and 0.80 with $[FD, SI, F_{cc}]$ or $[FD, C]$. Other combinations of the measures provided lower AUC.

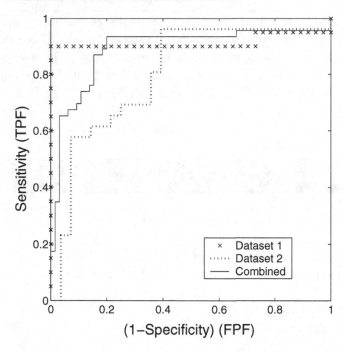

Figure 6.3: ROC curves indicating the classification performance of FD obtained using the ruler method applied to the 1D signatures of the first ("×"), second (dotted line), and combined (solid line) datasets. The AUC values are 0.91, 0.80, and 0.89, respectively. TPF = true-positive fraction; FPF = false-positive fraction. Reproduced, with kind permission from Springer Science+Business Media B. V., from R. M. Rangayyan and T. M. Nguyen, "Fractal analysis of contours of breast masses in mammograms," *Journal of Digital Imaging*, 20(3):223–237, 2007. © Springer.

- **Combined dataset**— The combined dataset gives a good representation of the combinations of common and atypical breast mass contours that are encountered in a clinical setting. In single-feature classification, SI provided the highest AUC of 0.90, followed by FD (0.89), F_{cc} (0.88), and C (0.87). In multiple-feature classification, the combination $[FD, F_{cc}]$ yielded the highest AUC of 0.93. Other combinations of features that yielded high AUC are: $[FD, F_{cc}, SI]$, $[SI, F_{cc}]$, and $[FD, FF, SI, F_{cc}, C]$, all with AUC = 0.92.

In general, the use of multiple shape measures could lead to more accurate pattern classification, as indicated by higher AUC, than the use of a single shape feature. By combining different shape features, the weaknesses of one shape feature may be compensated by the strengths of the other shape features. The combination of FD and F_{cc} consistently yielded the highest accuracy across both datasets and the combined dataset. The results indicate that FD can complement F_{cc} in the classification of atypical masses and tumors based upon their contours. Figure 6.4 shows ROC plots

Table 6.2: Comparison of AUC for various combinations of shape measures and FD obtained using the ruler method applied to 1D signatures. The Bayesian classifier was used for pattern classification with more than one feature. Reproduced, with kind permission from Springer Science+Business Media B. V., from R. M. Rangayyan and T. M. Nguyen, "Fractal analysis of contours of breast masses in mammograms," *Journal of Digital Imaging*, 20(3):223–237, 2007. © Springer.

Shape Features	Dataset 1	Dataset 2	Combined
FD, F_{cc}	0.99	0.82	0.93
FD, SI, F_{cc}	0.99	0.80	0.92
SI, F_{cc}	0.99	0.77	0.92
FD, SI	0.97	0.82	0.91
SI	0.93	0.77	0.90
FD	0.91	0.80	0.89
F_{cc}	0.99	0.77	0.88
C	0.97	0.72	0.87
FF	0.98	0.65	0.77

representing the classification of masses with FD obtained using the ruler method applied to 1D signatures, F_{cc}, and the two features combined, using the combined dataset.

In order to analyze the strengths and weaknesses of each shape feature, a sample classification experiment was carried out using the combined dataset. The prior probabilities of the benign and malignant classes estimated for the combined dataset (with 111 contours, 65 of which are related to benign masses and 46 of which are related to malignant tumors) were used to obtain the threshold to classify the masses. All of the shape features, except FF, were able to classify the typical CB masses and SM tumors accurately. However, most of the shape features were unable to classify correctly the atypical cases of SB masses and CM tumors. The best fraction of correctly classified CM tumors was achieved with FF, but FF was poor in classifying the CB masses and the SM tumors. Therefore, FF, when combined with other features, did not improve the classification accuracy.

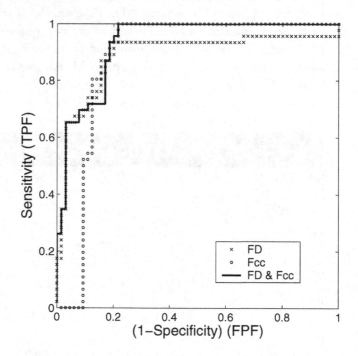

Figure 6.4: ROC curves representing the classification performance of FD obtained with the ruler method applied to 1D signatures ("×"), F_{cc} ("○"), and their combination (solid line) with the combined dataset. The AUC values are 0.89, 0.88, and 0.93, respectively. Reproduced, with kind permission from Springer Science+Business Media B. V., from R. M. Rangayyan and T. M. Nguyen, "Fractal analysis of contours of breast masses in mammograms," *Journal of Digital Imaging*, 20(3):223–237, 2007. © Springer.

The next highest fraction of correctly classified CM tumors was achieved by F_{cc}. The best fraction of correctly classified SB masses was achieved by FD. Both F_{cc} and FD were reasonably accurate in classifying the CB masses and SM tumors. F_{cc} was weak in classifying the SB masses, but effective in classifying the CM tumors. On the other hand, FD was weak in classifying the CM tumors, but was the best feature in classifying the SB masses. This indicates that FD and F_{cc} can compensate for each other's weaknesses. The combination of F_{cc} and FD provided the highest AUC of 0.93 with the combined dataset.

6.3   RESULTS OF ANALYSIS OF GRAY-SCALE VARIATION

Subtle textural differences have been observed between benign masses and malignant tumors, with the former being mostly homogeneous and the latter showing heterogeneous and nonuniform texture. This section presents the results of classification using FD and Haralick's texture measures computed using ROIs of breast masses (regions within their contours) and the ribbon regions of the breast masses (see Figures 3.1 to 3.6 for illustrations of ROIs and ribbons of breast masses).

6.3.1   FRACTAL ANALYSIS OF GRAY-SCALE VARIATION

The blanket method described in Section 4.3.1 to compute the gray-scale FD was applied to the ROIs and ribbons of breast masses. The classification results using ROC analysis and the p-values are presented in Table 6.3. For the combined dataset, the AUC obtained using the gray-scale FD computed from the ROIs and ribbons of breast masses were 0.59 and 0.67, respectively. The poor classification results (in terms of AUC values) may be explained by the small size of some of the ROIs and ribbons in the datasets. After a few levels of downsampling, many of the ROIs and ribbon regions contain very few pixels to derive any reliable measures. The narrowest ROI in the datasets is only about 100 pixels wide. Compared to the ROIs, the ribbons have even fewer pixels in their images. For this reason, only a small range of scale or pixel resolution of the ROIs and ribbons could be used to estimate a linear fit on the log–log curve to compute FD. At downsampled levels greater than 6×6 pixels square, the log–log curves for most of the cases exhibited no relationship between the log of the downsampling level and the log of the gray-scale surface area. Thus, the FD values derived from estimates of the gray-scale surface area at various levels of resolution of the ROIs and ribbons of breast masses are poor measures for the classification of breast masses, as indicated by the results obtained. Furthermore, all of the p-values shown in Table 6.3, except two, indicate that the difference between the mean FD of benign masses and the mean FD of malignant tumors is not statistically significant. The distributions of the benign and malignant classes are inseparable for these cases, and hence caused low AUC values.

Table 6.3: Results of classification (in terms of AUC) of ROIs and ribbons of breast masses using the blanket method to compute gray-scale FD. The corresponding p-values are shown in parentheses.

Mass Image	Dataset 1	Dataset 2	Combined
ROI	0.57 (p=0.44)	0.65 (p=0.23)	0.59 (p=0.20)
Ribbon	0.55 (p=0.76)	0.78 ($p < 0.01$)	0.67 (p=0.01)

6.3.2   HARALICK'S TEXTURE MEASURES

In order to study the effect of pixel size on texture analysis [19], the images in the combined dataset were filtered and downsampled to effective pixel size or resolution of 100, 200, 400, 600, 800, and 1000 μm. The downsampling process used Gaussian filters of standard deviation 1, 2, 4, 6, 8, and 10 pixels, and square mask (convolution matrix) sizes of 7, 13, 25, 37, 45, and 61 pixels, respectively, as antialiasing filters prior to decimation. The corresponding ribbons of masses were also downsampled to the same pixel resolution. The GLCMs were computed at each level of pixel resolution for the mass ROIs and ribbons, maintaining 256 levels of gray. Cooccurrences were combined for the four directions of $0°$, $45°$, $90°$, and $135°$ with unit pixel distance at each level of resolution. All of the 14 GLCM-based texture features were computed for each mass ROI and ribbon at each level of resolution, using the definitions of Haralick et al. [27], as described in Section 2.3.3.

Table 6.4 lists the classification accuracy, in terms of AUC, for each individual texture feature computed from the ROIs and ribbons of the 111 masses tested, at the various pixel sizes considered. A sliding threshold was applied to each feature to compute the measures required for ROC analysis; no trained classifier was used in this experiment. In classification experiments using each texture feature individually, the texture feature F_{10}, difference variance, gave the best classification results with a maximum AUC of 0.72 as well as high classification performance (AUC) across several levels of resolution. At certain levels of resolution, the texture features F_2, F_9, and F_{11} individually gave similar classification results, with AUC = 0.72.

The results of pattern classification with FLDA and the leave-one-out (LOO) method, listed in Table 6.5, indicate poor performance, with the maximal accuracy of AUC = 0.71, achieved individually by F_9 and F_{11}, and by the combination of all 14 features computed using ribbons of masses with the pixel size of 800 and 1000 μm. (Cases with the AUC under 0.5 are due to the effects of the LOO procedure, and indicate that reliable separation between the two classes is not feasible under the circumstances of the associated experiments.) When the AUC values across all of the FLDA experiments were considered, the texture features F_1, F_4, and F_8 individually gave the best overall performance. The results indicate that benign masses and malignant tumors are not linearly separable, in the space defined by the 14 texture features, to a high level of classification accuracy.

The values of AUC obtained using Haralick's texture features computed from the entire ROI and ribbons at several levels of pixel resolution with the Bayesian classifier are shown in Table 6.6. The classification experiment using the Bayesian classifier and LOO yielded higher values of AUC than FLDA, with the best performance of AUC = 0.75 given by the full set of 14 texture features computed using mass ribbons at the pixel resolution of 400 and 800 μm. Individually, the feature F_{11} gave the best performance with AUC = 0.70; considering the AUC across all the experiments, the texture feature F_8 gave the most consistent performance. The results indicate that the use of a nonlinear classifier with all of Haralick's 14 texture features together, computed using ribbons of pixels around the masses, can provide higher performance in discriminating between benign masses and malignant tumors than the use of individual texture features and linear classifiers. It should be noted that, in cases with poor separation (or complete overlap) between the features of the two classes, use of the LOO method resulted in AUC values less than 0.5.

Table 6.7 lists the mean and standard deviation values of AUC obtained with 100 repetitions of 50% random splitting of the sets of 14 texture features into the training and testing parts of the Bayesian classifier. The ribbons of masses were used to compute the texture features. The highest

Table 6.4: The classification accuracy, measured in terms of the AUC, for Haralick's texture features computed from the ROIs and ribbons of 111 masses at several levels of pixel resolution (in μm). A sliding threshold was applied to each feature; no trained classifier was used in this experiment. Reproduced, with kind permission from Springer Science+Business Media B. V., from R. M. Rangayyan, T. M. Nguyen, F. J. Ayres, and A. K. Nandi, "Effect of pixel resolution on texture features of breast masses in mammograms," *Journal of Digital Imaging*, 23(5):547–553, 2010. © Springer.

Region	Pixel	F1	F2	F3	F4	F5	F6	F7	F8	F9	F10	F11	F12	F13	F14
ROI	50	0.63	0.54	0.68	0.70	0.51	0.51	0.70	0.66	0.64	0.54	0.53	0.65	0.69	0.67
	100	0.69	0.66	0.57	0.70	0.65	0.51	0.70	0.66	0.66	0.67	0.66	0.51	0.60	0.57
	200	0.71	0.71	0.51	0.70	0.70	0.50	0.70	0.69	0.72	0.71	0.71	0.58	0.54	0.51
	400	0.71	0.71	0.55	0.69	0.71	0.51	0.69	0.70	0.71	0.71	0.72	0.51	0.59	0.56
	600	0.68	0.72	0.55	0.69	0.70	0.51	0.69	0.67	0.66	0.71	0.72	0.61	0.65	0.53
	800	0.65	0.71	0.56	0.69	0.71	0.51	0.68	0.66	0.62	0.72	0.72	0.63	0.65	0.53
	1000	0.60	0.72	0.56	0.69	0.70	0.51	0.63	0.63	0.57	0.72	0.71	0.63	0.64	0.51
Ribbon	50	0.59	0.51	0.65	0.64	0.54	0.58	0.64	0.66	0.59	0.53	0.52	0.64	0.68	0.65
	100	0.65	0.67	0.52	0.64	0.63	0.58	0.64	0.66	0.66	0.68	0.65	0.51	0.58	0.54
	200	0.69	0.71	0.56	0.64	0.68	0.58	0.64	0.66	0.69	0.72	0.69	0.59	0.51	0.53
	400	0.69	0.71	0.58	0.64	0.70	0.58	0.63	0.65	0.69	0.72	0.70	0.56	0.53	0.57
	600	0.68	0.70	0.58	0.63	0.70	0.58	0.62	0.64	0.67	0.70	0.70	0.52	0.58	0.57
	800	0.66	0.69	0.58	0.63	0.69	0.57	0.62	0.63	0.63	0.69	0.69	0.56	0.59	0.57
	1000	0.62	0.68	0.58	0.63	0.69	0.57	0.61	0.61	0.60	0.68	0.69	0.56	0.58	0.56

Table 6.5: The classification accuracy, measured in terms of the AUC, for Haralick's texture features computed from the ROIs and ribbons of 111 masses at several levels of pixel resolution (in μm). FLDA was used, with the LOO method for cross-validation. Reproduced, with kind permission from Springer Science+Business Media B. V., from R. M. Rangayyan, T. M. Nguyen, F. J. Ayres, and A. K. Nandi, "Effect of pixel resolution on texture features of breast masses in mammograms," *Journal of Digital Imaging*, 23(5):547–553, 2010. © Springer.

Region	Pixel	All 14	F1	F2	F3	F4	F5	F6	F7	F8	F9	F10	F11	F12	F13	F14
ROI	50	0.64	0.61	0.33	0.67	0.68	0.35	0.42	0.68	0.69	0.62	0.39	0.01	0.64	0.68	0.65
	100	0.63	0.67	0.61	0.54	0.68	0.63	0.41	0.68	0.69	0.68	0.62	0.64	0.01	0.57	0.54
	200	0.60	0.70	0.68	0.44	0.68	0.68	0.39	0.68	0.69	0.71	0.67	0.69	0.56	0.48	0.34
	400	0.61	0.70	0.68	0.52	0.68	0.70	0.31	0.67	0.67	0.69	0.69	0.70	0.36	0.56	0.52
	600	0.66	0.67	0.68	0.52	0.67	0.69	0.27	0.67	0.66	0.65	0.69	0.71	0.59	0.63	0.49
	800	0.59	0.64	0.69	0.54	0.67	0.70	0.24	0.66	0.64	0.61	0.69	0.71	0.61	0.63	0.48
	1000	0.60	0.58	0.69	0.53	0.67	0.69	0.21	0.66	0.61	0.55	0.69	0.70	0.62	0.61	0.14
Ribbon	50	0.59	0.57	0.00	0.63	0.63	0.52	0.55	0.63	0.65	0.57	0.33	0.48	0.63	0.65	0.62
	100	0.66	0.64	0.61	0.43	0.63	0.62	0.55	0.63	0.65	0.64	0.63	0.63	0.38	0.54	0.46
	200	0.68	0.67	0.68	0.52	0.63	0.67	0.55	0.62	0.64	0.67	0.69	0.68	0.57	0.29	0.49
	400	0.68	0.68	0.68	0.55	0.62	0.68	0.55	0.61	0.64	0.68	0.69	0.69	0.54	0.44	0.54
	600	0.69	0.67	0.67	0.56	0.61	0.68	0.55	0.61	0.63	0.66	0.67	0.68	0.42	0.52	0.55
	800	0.71	0.64	0.66	0.55	0.61	0.68	0.55	0.60	0.62	0.62	0.66	0.68	0.53	0.52	0.55
	1000	0.71	0.61	0.66	0.55	0.60	0.68	0.55	0.59	0.59	0.58	0.65	0.67	0.53	0.46	0.54

Table 6.6: The classification accuracy, measured in terms of the AUC, for Haralick's texture features computed from the ROIs and ribbons of 111 masses at several levels of pixel resolution (in μm). The Bayesian classifier was used, with the LOO method for cross-validation. Reproduced, with kind permission from Springer Science+Business Media B. V., from R. M. Rangayyan, T. M. Nguyen, F. J. Ayres, and A. K. Nandi, "Effect of pixel resolution on texture features of breast masses in mammograms," *Journal of Digital Imaging*, 23(5):547–553, 2010. © Springer.

Region	Pixel	All 14	F1	F2	F3	F4	F5	F6	F7	F8	F9	F10	F11	F12	F13	F14
ROI	50	0.65	0.59	0.62	0.62	0.61	0.22	0.41	0.60	0.69	0.58	0.26	0.00	0.60	0.58	0.60
	100	0.65	0.62	0.45	0.40	0.59	0.62	0.42	0.59	0.69	0.67	0.43	0.63	0.42	0.55	0.41
	200	0.66	0.63	0.53	0.38	0.59	0.68	0.42	0.58	0.68	0.70	0.53	0.68	0.55	0.41	0.18
	400	0.69	0.62	0.55	0.51	0.57	0.70	0.43	0.56	0.67	0.69	0.54	0.70	0.23	0.47	0.44
	600	0.70	0.59	0.54	0.50	0.57	0.69	0.44	0.55	0.65	0.65	0.53	0.70	0.58	0.50	0.40
	800	0.67	0.53	0.58	0.55	0.56	0.69	0.43	0.53	0.64	0.60	0.58	0.70	0.61	0.53	0.36
	1000	0.67	0.44	0.57	0.56	0.57	0.68	0.44	0.53	0.60	0.48	0.58	0.70	0.61	0.52	0.44
Ribbon	50	0.66	0.54	0.24	0.61	0.62	0.46	0.47	0.62	0.63	0.54	0.24	0.36	0.59	0.60	0.60
	100	0.71	0.61	0.46	0.26	0.62	0.59	0.47	0.62	0.63	0.62	0.45	0.60	0.28	0.45	0.32
	200	0.73	0.64	0.56	0.42	0.61	0.66	0.47	0.61	0.63	0.66	0.57	0.67	0.56	0.09	0.37
	400	0.75	0.65	0.55	0.49	0.60	0.68	0.47	0.60	0.62	0.67	0.56	0.68	0.51	0.32	0.48
	600	0.74	0.63	0.61	0.51	0.59	0.67	0.47	0.59	0.61	0.65	0.56	0.68	0.26	0.41	0.50
	800	0.75	0.55	0.62	0.54	0.59	0.67	0.46	0.58	0.60	0.61	0.60	0.67	0.45	0.41	0.54
	1000	0.70	0.50	0.62	0.55	0.58	0.67	0.46	0.56	0.58	0.55	0.62	0.67	0.46	0.43	0.54

mean AUC of 0.6464 was obtained at 200 μm per pixel. The results of t-tests indicated that the AUC values obtained at 200 μm per pixel are greater than those obtained at all other pixel sizes with high statistical significance ($p < 0.01$), except 100 μm. The AUC values at 800 or 1000 μm per pixel are lower than those at any other pixel size with high statistical significance ($p < 0.01$); however, the AUC values at 800 μm vs. those at 1000 μm per pixel have no statistically significant difference. The AUC values for 50 μm per pixel are lower than those for 100 and 200 μm per pixel, and higher than those for 800 and 1000 μm per pixel, with high statistical significance ($p < 0.01$).

Table 6.7: Mean and standard deviation of the classification accuracy, measured in terms of the AUC, for Haralick's texture features computed from the ribbons of 111 masses at several levels of pixel resolution. Reproduced, with kind permission from Springer Science+Business Media B. V., from R. M. Rangayyan, T. M. Nguyen, F. J. Ayres, and A. K. Nandi, "Effect of pixel resolution on texture features of breast masses in mammograms," *Journal of Digital Imaging*, 23(5):547–553, 2010. © Springer.

	Pixel resolution (μm)						
	50	100	200	400	600	800	1000
Mean AUC	0.59	0.63	0.65	0.61	0.61	0.56	0.55
Standard deviation	0.07	0.06	0.07	0.07	0.08	0.07	0.07

6.4   DISCUSSION

Fractal analysis is useful for characterization and quantitative representation of the complexity of contours of breast masses. FD was observed to complement the shape feature F_{cc} in the classification of breast masses as benign or malignant based on the shape of the breast masses.

Although FD performed exceptionally well as a shape measure, it did not perform as well as a texture measure for breast masses. Although some studies have reported that fractal analysis of gray-scale variability can be used to discriminate breast lesions as benign and malignant, with one study by Chen et al. [91] achieving a classification performance of AUC = 0.88 (using ultrasound imaging), the studies conducted in the present work did not achieve such high results. Haralick's texture measures performed better than FD in the analysis of gray-scale variability in ROIs and ribbons of breast masses. Haralick's texture features yielded higher classification accuracy when computed from the ribbons of masses than from the entire mass ROIs; see also Mudigonda et al. [22, 24]. Furthermore, in the study on the effect of pixel size on texture measures computed based on the GLCM, it was observed that the texture features computed with the pixel size of 200 μm provided the highest classification results (based on AUC values) as compared to all of the other pixel sizes tested.

Lee et al. [92] conducted a study on the effect of the number of gray levels used on texture analysis of mammograms. The best results were obtained using features derived from a combination of three quantization levels.

A limitation of the present study is that the contours used were drawn by hand on mammograms (by an expert radiologist specialized in mammography). The methods for shape and texture analysis need to be tested with contours obtained automatically by image processing methods for the detection and delineation of masses in mammograms [22, 23]. It is worth noting that, in a study by Sahiner et al. [23], in which the classification performance of several shape and texture measures was compared with a dataset of automatically extracted regions corresponding to 122 benign breast masses and 127 malignant tumors, FF was found to give the best individual performance with AUC = 0.82. This result not only indicates the importance of shape in the analysis of breast masses, but also that shape factors computed from automatically extracted contours can yield good results in discriminating between benign masses and malignant tumors. It would also be desirable to test the methods for both shape and gray-scale analysis with contours of masses drawn by several radiologists in order to assess the effects of interobserver variability.

Another limitation of the present study is caused by the size of the images containing masses, with the smallest image being only a few hundred pixels wide. Downsampling of such images resulted in even smaller images with insufficient numbers of pixels to permit the derivation of reliable GLCMs and texture features as well as gray-scale FD. This limitation affected the results presented in Sections 6.3.1 and 6.3.2. Other methods to compute FD and texture measures are desirable to overcome this limitation.

CHAPTER 7

Concluding Remarks

We presented the results of an investigation of fractal analysis for the classification of breast masses on digitized mammograms based on their shape and gray-scale texture [18, 19]. The present study also compared the FD of breast masses against and in conjunction with combinations of shape features proposed in previous related studies by Rangayyan et al. [25, 26].

In shape analysis of breast masses, the experimental results show statistically highly significant differences in FD between malignant tumors and benign masses. The FD of benign masses is, in general, lower than the FD of malignant tumors. The results obtained using the ruler method suggest that the 2D contour representation can provide a slightly more accurate classification than the 1D signature representation, whereas the results obtained using the box-counting method show that the 1D signature representation provides a slightly more accurate classification than the 2D contour representation. The distinction between the performance of the 2D contour and 1D signature representations is not consistent between the box-counting method and the ruler method. However, the combined dataset confirms that the 1D signature representation consistently yields more accurate classification than the 2D contour representation. The ruler method provided a slightly more accurate classification than the box-counting method. Furthermore, use of the 1D signature facilitates application of the PSA method based on the fBm model to derive FD using a different approach.

With a dataset including 111 contours of a combination of typical and atypical masses and tumors, FD gave an accuracy comparable to that of other shape features such as F_{cc} and SI. Furthermore, FD was able to classify atypical benign masses that are spiculated more accurately than the other shape features studied; it was, however, less accurate in the classification of malignant tumors that are circumscribed. When FD was combined with other shape features, specifically F_{cc} and SI, the classification accuracy was improved. The results obtained indicate that FD can serve as a useful shape feature, by itself or in conjunction with other shape features, in the classification of breast masses and the diagnosis of breast cancer.

In fractal analysis of gray-scale variation in breast masses, the experimental results show that the FD of benign masses is, in general, greater than the FD of malignant tumors. This could indicate that malignant tumors disrupt the self-similarity present in normal breast parenchymal patterns to a larger extent than benign masses. However, the results indicate poorer classification capability of gray-scale FD as compared to shape-based FD and other shape factors.

In statistical analysis of masses using Haralick's GLCM-based features, the classification results indicate that the features are affected by the pixel size or resolution of the image. The ribbons or margins of masses at pixel resolution finer than 200 μm or coarser than 800 μm per pixel are unsuitable for use in the classification of mammographic masses. The use of coarse resolution (large pixel size) via downsampling of the original image could lead to the number of pixels available, in the case of small masses, being too low to permit the derivation of reliable GLCMs and texture features. The pixel resolution of 200 μm is most suitable for the computation of GLCM-based texture features using ribbons or margins of masses. The texture features computed using the ribbons of masses yielded higher classification accuracy than the same features computed using the entire mass ROIs. Regardless, the highest classification accuracy obtained using Haralick's texture features is lower than that with any of the shape features.

The classification performance of shape features in the present study is higher than that of the shape features used in the study of Sahiner et al. [23], whereas the classification performance of the texture measures is comparable.

Although the present study has demonstrated the success of shape analysis in discriminating between benign masses and malignant tumors, it should be noted that shape analysis requires accurate contours. The contours used in the present study were drawn on mammographic images by a highly experienced radiologist. It is desirable to test fractal analysis and other shape factors on automatically segmented contours of breast masses.

The present study tested manually selected combinations of features in pattern classification experiments. It is desirable to apply feature selection methods to determine optimal combinations of shape and texture features for improved classification of breast masses. Further work with advanced classifiers could lead to higher classification accuracies than those reported in the present study [93, 94, 95].

It would be appropriate and timely to test the methods proposed in this book with direct digital mammograms and breast tomosynthesis images acquired with the latest imaging technology. The methods could be used in conjunction with procedures for the detection and analysis of masses for CAD of breast cancer.

Bibliography

[1] A. K. Hackshaw and E. A. Paul. Breast self-examination and death from breast cancer: a meta-analysis. *British Journal of Cancer*, 88:1047–1053, 2003. DOI: 10.1038/sj.bjc.6600847 Cited on page(s) 1

[2] K. Doi. Diagnostic imaging over the last 50 years: research and development in medical imaging science and technology. *Physics in Medicine and Biology*, 51:R5–R27, 2006. DOI: 10.1088/0031-9155/51/13/R02 Cited on page(s) 1

[3] M. J. Homer. *Mammographic Interpretation: A Practical Approach*. McGraw-Hill, Boston, MA, 2nd edition, 1997. Cited on page(s) 1, 2, 8, 16

[4] Canadian Cancer Society, http://www.cancer.ca/Canada-wide/Prevention/Getting%20checked/Breast%20cancer%20NEW.aspx?sc_lang=en. *Breast cancer screening guidelines*, accessed September 2012. Cited on page(s) 1

[5] R. E. Bird, T. W. Wallace, and B. C. Yankaskas. Analysis of cancers missed at screening mammography. *Radiology*, 184:613–617, 1992. Cited on page(s) 2

[6] R. G. Blanks, M. G. Wallis, and S. M. Moss. A comparison of cancer detection rates achieved by breast cancer screening programmes by number of readers, for one and two view mammography: results from the UK National Health Service breast screening programme. *Journal of Medical Screening*, 5:195–201, 1998. DOI: 10.1136/jms.5.4.195 Cited on page(s) 2

[7] S. C. Harvey, B. Geller, R. G. Oppenheimer, M. Pinet, L. Riddell, and B. Garra. Increase in cancer detection and recall rates with independent double interpretation of screening mammography. *American Journal of Roentgenology*, 180:1461–1467, 2003. Cited on page(s) 2

[8] S. Ciatto, D. Ambrogetti, R. Bonardi, S. Catarzi, G. Risso, M. Rosselli Del Turco, and P. Mantellini. Second reading of screening mammograms increases cancer detection and recall rates. Results in the Florence screening programme. *Journal of Medical Screening*, 12:103–106, 2005. DOI: 10.1258/0969141053908285 Cited on page(s) 2

[9] J. G. Elmore and R. J. Brenner. The more eyes, the better to see? From double to quadruple reading of screening mammograms. *Journal of the National Cancer Institute*, 99(15):1141–1143, August 2007. DOI: 10.1093/jnci/djm079 Cited on page(s) 2

[10] E. D. Pisano, C. Gatsonis, E. Hendrick, M. Yaffe, J. K. Baum, S. Acharyya, E. F. Conant, L. L. Fajardo, L. Bassett, C. D'Orsi, R. Jong, and M. Rebner. Diagnostic performance of digital versus film mammography for breast-cancer screening. *New England Journal of Medicine*, 353(17):1773–1783, October 2005. DOI: 10.1056/NEJMoa052911 Cited on page(s) 3

[11] Alberta Health Services, http://www.albertahealthservices.ca/services.asp?pid=service&rid=1002353. *Screen Test and the Alberta Breast Cancer Screening Program*, accessed September 2012. Cited on page(s) 3, 25

[12] L. W. Bassett, B. S. Monsees, R. A. Smith, L. Wang, P. Hooshi, D. M. Farria, J. W. Sayre, S. A. Feig, and V. P. Jackson. Survey of radiology residents: breast imaging training and attitude. *Radiology*, 227:862–869, 2003. DOI: 10.1148/radiol.2273020046 Cited on page(s) 3

[13] K. Doi. Computer-aided diagnosis in medical imaging: historical review, current status and future potential. *Computerized Medical Imaging and Graphics*, 31:198–211, 2007. DOI: 10.1016/j.compmedimag.2007.02.002 Cited on page(s) 3

[14] T. W. Freer and M. J. Ulissey. Screening mammography with computer-aided detection: prospective study of 12,860 patients in a community breast center. *Radiology*, 220:781–786, August 2001. DOI: 10.1148/radiol.2203001282 Cited on page(s) 4

[15] M. Gromet. Comparison of computer-aided detection to double reading of screening mammograms: review of 231,221 mammograms. *American Journal of Roentgenology*, 190:854–859, 2008. DOI: 10.2214/AJR.07.2812 Cited on page(s) 4

[16] F. J. Gilbert, S. M. Astley, M. G. C. Gillian, O. F. Agbaje, M. G. Wallis, J. James, C. R. M. Boggis, and S. W. Duffy. Single reading with computer-aided detection for screening mammography. *New England Journal of Medicine*, 359(16):1675–1684, October 2008. DOI: 10.1056/NEJMoa0803545 Cited on page(s) 4

[17] J. J. Fenton, S. H. Taplin, P. A. Carney, L. Abraham, E. A. Sickles, C. D'Orsi, E. A. Berns, G. Cutter, R. E. Hendrick, W. E. Barlow, and J. G. Elmore. Influence of computer-aided detection on performance of screening mammography. *New England Journal of Medicine*, 356(14):1399–1409, 2007. DOI: 10.1056/NEJMoa066099 Cited on page(s) 5

[18] R. M. Rangayyan and T. M. Nguyen. Fractal analysis of contours of breast masses in mammograms. *Journal of Digital Imaging*, 20(3):223–237, September 2007. DOI: 10.1007/s10278-006-0860-9 Cited on page(s) 5, 8, 29, 38, 41, 72, 85

[19] R. M. Rangayyan, T. M. Nguyen, F. J. Ayres, and A. K. Nandi. Effect of pixel resolution on texture features of breast masses in mammograms. *Journal of Digital Imaging*, 23(5):547–553, October 2010. DOI: 10.1007/s10278-009-9238-0 Cited on page(s) 5, 29, 77, 85

[20] B. B. Mandelbrot. *The Fractal Geometry of Nature*. W. H. Freeman and Company, San Francisco, CA, 1983. Cited on page(s) 5, 31, 32, 33, 40

[21] H. Alto, R. M. Rangayyan, and J. E. L. Desautels. Content-based retrieval and analysis of mammographic masses. *Journal of Electronic Imaging*, 14(2):023016:1–17, 2005. DOI: 10.1117/1.1902996 Cited on page(s) 5, 8, 15, 16, 24, 25, 29

[22] N. R. Mudigonda, R. M. Rangayyan, and J. E. L. Desautels. Detection of breast masses in mammograms by density slicing and texture flow-field analysis. *IEEE Transactions on Medical Imaging*, 20(12):1215–1227, 2001. DOI: 10.1109/42.974917 Cited on page(s) 5, 9, 16, 24, 29, 83

[23] B. S. Sahiner, H. P. Chan, N. Petrick, M. A. Helvie, and L. M. Hadjiiski. Improvement of mammographic mass characterization using spiculation measures and morphological features. *Medical Physics*, 28(7):1455–1465, 2001. DOI: 10.1118/1.1381548 Cited on page(s) 5, 9, 16, 24, 83, 86

[24] N. R. Mudigonda, R. M. Rangayyan, and J. E. L. Desautels. Gradient and texture analysis for the classification of mammographic masses. *IEEE Transactions on Medical Imaging*, 19(10):1032–1043, 2000. DOI: 10.1109/42.887618 Cited on page(s) 5, 16, 24, 29, 83

[25] R. M. Rangayyan, N. M. El-Faramawy, J. E. L. Desautels, and O. A. Alim. Measures of acutance and shape for classification of breast tumors. *IEEE Transactions on Medical Imaging*, 16(6):799–810, 1997. DOI: 10.1109/42.650876 Cited on page(s) 5, 8, 14, 15, 16, 24, 25, 29, 85

[26] R. M. Rangayyan, N. R. Mudigonda, and J. E. L. Desautels. Boundary modelling and shape analysis methods for classification of mammographic masses. *Medical and Biological Engineering and Computing*, 38:487–496, 2000. DOI: 10.1007/BF02345742 Cited on page(s) 5, 8, 14, 15, 16, 25, 29, 85

[27] R. M. Haralick, K. Shanmugam, and I. Dinstein. Textural features for image classification. *IEEE Transactions on Systems, Man, Cybernetics*, SMC-3(6):610–622, 1973. DOI: 10.1109/TSMC.1973.4309314 Cited on page(s) 5, 16, 20, 24, 78

[28] R. M. Rangayyan. *Biomedical Image Analysis*. CRC Press, Boca Raton, FL, 2005. Cited on page(s) 7, 14, 24

[29] R. M. Rangayyan, F. J. Ayres, and J. E. L. Desautels. A review of computer-aided diagnosis of breast cancer: Toward the detection of subtle signs. *Journal of the Franklin Institute*, 344:312–348, 2007. DOI: 10.1016/j.jfranklin.2006.09.003 Cited on page(s) 7

[30] J. Tang, R. M. Rangayyan, J. Xu, I. E. Naqa, and Y. Yang. Computer-aided detection and diagnosis of breast cancer with mammography: Recent advances. *IEEE Transactions on Information*

Technology in Biomedicine, 13(2):236–251, March 2009. DOI: 10.1109/TITB.2008.2009441 Cited on page(s) 7

[31] American College of Radiology, Reston, VA. *Illustrated Breast Imaging Reporting and Data System (BI-RADS*^{TM}*)*, third edition, 1998. Cited on page(s) 8, 16

[32] T. Matsubara, H. Fujita, S. Kasai, M. Goto, Y. Tani, T. Hara, and T. Endo. Development of new schemes for detection and analysis of mammographic masses. In *Proceedings of the 1997 IASTED International Conference on Intelligent Information Systems (IIS'97)*, pages 63–66, Grand Bahama Island, Bahamas, December 1997. DOI: 10.1109/IIS.1997.645180 Cited on page(s) 8, 40

[33] B. S. Sahiner, H. P. Chan, N. Petrick, M. A. Helvie, and M. M. Goodsitt. Computerized characterization of masses on mammograms: The rubber band straightening transform and texture analysis. *Medical Physics*, 25(4):516–526, 1998. DOI: 10.1118/1.598228 Cited on page(s) 9

[34] L. Zheng and A. K. Chan. An artificial intelligent algorithm for tumor detection in screening mammogram. *IEEE Transactions on Medical Imaging*, 20(7):559–567, 2001. DOI: 10.1109/42.932741 Cited on page(s) 9, 40

[35] A. Rojas Dominguez and A. K. Nandi. Toward breast cancer diagnosis based on automated segmentation of masses in mammograms. *Pattern Recognition*, 42:1138–1148, 2009. DOI: 10.1016/j.patcog.2008.08.006 Cited on page(s) 10

[36] S. Pohlman, K. A. Powell, N. A. Obuchowski, W. A. Chilcote, and S. Grundfest-Broniatowski. Quantitative classification of breast tumors in digitized mammograms. *Medical Physics*, 23(8):1337–1345, 1996. DOI: 10.1118/1.597707 Cited on page(s) 11, 40

[37] L. Shen, R. M. Rangayyan, and J. E. L. Desautels. Detection and classification of mammographic calcifications. *International Journal of Pattern Recognition and Artificial Intelligence*, 7(6):1403–1416, 1993. DOI: 10.1142/S0218001493000686 Cited on page(s) 14, 15

[38] L. Shen, R. M. Rangayyan, and J. E. L. Desautels. Application of shape analysis to mammographic calcifications. *IEEE Transactions on Medical Imaging*, 13(2):263–274, 1994. DOI: 10.1109/42.293919 Cited on page(s) 15

[39] H. Alto, R. M. Rangayyan, R. B. Paranjape, J. E. L. Desautels, and H. Bryant. An indexed atlas of digital mammograms for computer-aided diagnosis of breast cancer. *Annales des Télécommunications*, 58(5-6):820–835, 2003. DOI: 10.1007/BF03001532 Cited on page(s) 25

[40] The Mammographic Image Analysis Society digital mammogram database. http://peipa.essex.ac.uk/info/mias.html, accessed September 2012. Cited on page(s) 25

[41] J. Suckling, J. Parker, D. R. Dance, S. Astley, I. Hutt, C. R. M. Boggis, I. Ricketts, E. Stamatakis, N. Cerneaz, S. L. Kok, P. Taylor, D. Betal, and J. Savage. The Mammographic Image Analysis Society digital mammogram database. In A. G. Gale, S. M. Astley, D. R. Dance, and A. Y. Cairns, editors, *Proceedings of the 2nd International Workshop on Digital Mammography*, volume 1069 of *Excerpta Medica International Congress Series*, pages 375–378, York, UK, July 1994. Cited on page(s) 25

[42] H. O. Peitgen, H. Jürgens, and D. Saupe. *Chaos and Fractals: New Frontiers of Science*. Springer, New York, NY, 2004. Cited on page(s) 31, 32, 33, 35, 40, 42, 44, 51, 53

[43] S. H. Liu. Formation and anomalous properties of fractals. *IEEE Engineering in Medicine and Biology Magazine*, 11(2):28–39, June 1992. DOI: 10.1109/51.139034 Cited on page(s) 31, 32, 33, 40

[44] W. Deering and B. J. West. Fractal physiology. *IEEE Engineering in Medicine and Biology Magazine*, 11(2):40–46, June 1992. DOI: 10.1109/51.139035 Cited on page(s) 31, 37, 40

[45] H. E. Schepers, J. H. G. M. van Beek, and J. B. Bassingthwaighte. Four methods to estimate the fractal dimension from self-affine signals. *IEEE Engineering in Medicine and Biology Magazine*, 11(2):57–64, June 1992. DOI: 10.1109/51.139038 Cited on page(s) 31, 40

[46] C. Fortin, R. Kumaresan, W. Ohley, and S. Hoefer. Fractal dimension in the analysis of medical images. *IEEE Engineering in Medicine and Biology Magazine*, 11(2):65–71, June 1992. DOI: 10.1109/51.139039 Cited on page(s) 31, 32, 40

[47] A. L. Goldberger, D. R. Rigney, and B. J. West. Chaos and fractals in human physiology. *Scientific American*, 262:42–49, February 1990. DOI: 10.1038/scientificamerican0290-42 Cited on page(s) 31, 40

[48] R. F. Voss. Fractals in nature: From characterization to simulation. In H. O. Peitgen and D. Saupe, editors, *The Science of Fractal Images*, pages 21–69. Springer-Verlag, New York, NY, 1988. Cited on page(s) 32, 33, 40, 44, 48, 53

[49] D. Saupe. Algorithms for random fractals. In H. O. Peitgen and D. Saupe, editors, *The Science of Fractal Images*, pages 71–113. Springer-Verlag, New York, NY, 1988. Cited on page(s) 32, 48, 53

[50] R. M. Rangayyan and F. Oloumi. Fractal analysis and classification of breast masses using the power spectra of signatures of contours. *Journal of Electronic Imaging*, 21(2):023018:1–9, 2012. DOI: 10.1117/1.JEI.21.2.023018 Cited on page(s) 33, 44, 48, 51, 52

[51] J. A. Provine and R. M. Rangayyan. Lossless compression of Peanoscanned images. *Journal of Electronic Imaging*, 3(2):176–181, 1994. DOI: 10.1117/12.171928 Cited on page(s) 36

[52] G. Liew, J. J. Wang, N. Cheung, Y. P. Zhang, W. Hsu, M. L. Lee, P. Mitchell, G. Tikellis, B. Taylor, and T. Y. Wong. The retinal vasculature as a fractal: Methodology, reliability, and relationship to blood pressure. *Ophthalmology*, 115:1951–1956, 2008. DOI: 10.1016/j.ophtha.2008.05.029 Cited on page(s) 36

[53] T. Stŏsić and B. D. Stŏsić. Multifractal analysis of human retinal vessels. *IEEE Transactions on Medical Imaging*, 25(8):1101–1107, 2006. DOI: 10.1109/TMI.2006.879316 Cited on page(s) 36

[54] S. Kyriacos, F. Nekka, L. Cartilier, and P. Vico. Insights into the formation process of the retinal vasculature. *Fractals*, 5(4):615–624, 1997. DOI: 10.1142/S0218348X97000498 Cited on page(s) 36

[55] T. J. MacGillivray, N. Patton, F. N. Doubal, C. Graham, and J. M. Wardlaw. Fractal analysis of the retinal vascular network in fundus images. In *Proceedings of the 29th Annual International Conference of the IEEE Engineering in Medicine and Biology Society*, pages 6455–6458, Lyon, France, August 23-26 2007. IEEE. DOI: 10.1109/IEMBS.2007.4353837 Cited on page(s) 36

[56] A. L. Goldberger. Fractal mechanisms in the electrophysiology of the heart. *IEEE Engineering in Medicine and Biology Magazine*, 11:47–52, 1992. DOI: 10.1109/51.139036 Cited on page(s) 37

[57] A. L. Goldberger, V. Bhargava, B. J. West, and A. J. Mandell. On a mechanism of cardiac electrical stability: The fractal hypothesis. *Biophysical Journal*, 48:525–528, 1985. DOI: 10.1016/S0006-3495(85)83808-X Cited on page(s) 37

[58] S. Abboud, O. Berenfeld, and D. Sadeh. Simulation of high-resolution QRS complex using a ventricular model with a fractal conduction system. *Circulation Research*, 68:1751–1760, 1991. DOI: 10.1161/01.RES.68.6.1751 Cited on page(s) 37

[59] A. L. Goldberger and B. J. West. Fractals in physiology and medicine. *The Yale Journal of Biology and Medicine*, 60:421–435, 1987. Cited on page(s) 37, 38

[60] R. M. Rangayyan, F. Oloumi, Y. F. Wu, and S. X. Cai. Fractal analysis of knee-joint vibroarthrographic signals via power spectral analysis. *Biomedical Signal Processing and Control*, page in press, 2012. DOI: 10.1016/j.bspc.2012.05.004 Cited on page(s) 38, 44

[61] Y. Gazit, J. W. Baish, N. Safabakhsh, M. Leunig, L. T. Baxter, and R. K. Jain. Fractal characteristics of tumor vascular architecture during tumor growth and regression. *Microcirculation*, 4(4):395–402, 1997. DOI: 10.3109/10739689709146803 Cited on page(s) 38

[62] P. Dey and S. K. Mohanty. Fractal dimensions of breast lesions on cytology smears. *Diagnostic Cytopathology*, 29(2):85–86, 2003. DOI: 10.1002/dc.10324 Cited on page(s) 40

[63] Q. Guo, J. Shao, and V. F. Ruiz. Characterization and classification of tumor lesions using computerized fractal-based texture analysis and support vector machines in digital mammograms. *International Journal of Computer Assisted Radiology and Surgery*, 4(1):11–25, January 2009. DOI: 10.1007/s11548-008-0276-8 Cited on page(s) 40

[64] C. B. Caldwell, S. J. Stapleton, D. W. Holdsworth, R. A. Jong, W. J. Weiser, G. Cooke, and M. J. Yaffe. Characterization of mammographic parenchymal pattern by fractal dimension. *Physics in Medicine and Biology*, 35(2):235–247, 1990. DOI: 10.1088/0031-9155/35/2/004 Cited on page(s) 40, 51, 52

[65] J. W. Byng, N. F. Boyd, E. Fishell, R. A. Jong, and M. J. Yaffe. Automated analysis of mammographic densities. *Physics in Medicine and Biology*, 41:909–923, 1996. DOI: 10.1088/0031-9155/41/5/007 Cited on page(s) 40, 41, 51, 52

[66] R. M. Rangayyan, S. Prajna, F. J. Ayres, and J. E. L. Desautels. Detection of architectural distortion in mammograms acquired prior to the detection of breast cancer using Gabor filters, phase portraits, fractal dimension, and texture analysis. *International Journal of Computer Assisted Radiology and Surgery*, 2(6):347–361, April 2008. DOI: 10.1007/s11548-007-0143-z Cited on page(s) 41, 44, 52

[67] R. M. Rangayyan, S. Banik, and J. E. L. Desautels. Computer-aided detection of architectural distortion in prior mammograms of interval cancer. *Journal of Digital Imaging*, 23(5):611–631, October 2010. DOI: 10.1007/s10278-009-9257-x Cited on page(s) 41, 44, 52

[68] S. Banik, R. M. Rangayyan, and J. E. L. Desautels. Detection of architectural distortion in prior mammograms. *IEEE Transactions on Medical Imaging*, 30(2):279–294, February 2011. DOI: 10.1109/TMI.2010.2076828 Cited on page(s) 41, 44, 52, 53

[69] H. Li, M. L. Giger, O. I. Olopade, and L. Lan. Fractal analysis of mammographic parenchymal patterns in breast cancer risk assessment. *Academic Radiology*, 14:513–521, 2007. DOI: 10.1016/j.acra.2007.02.003 Cited on page(s) 41

[70] H. Li, M. L. Giger, O. I. Olopade, and M. R. Chinander. Power spectral analysis of mammographic parenchymal patterns for breast cancer risk assessment. *Journal of Digital Imaging*, 21(2):145–152, June 2008. DOI: 10.1007/s10278-007-9093-9 Cited on page(s) 41

[71] K. M. Iftekharuddin, W. Jia, and R. Marsh. Fractal analysis of tumor in brain MR images. *Machine Vision and Applications*, 13:352–362, 2003. DOI: 10.1007/s00138-002-0087-9 Cited on page(s) 41

[72] C. M. Wu, Y. C. Chen, and K. S. Hsieh. Texture features for classification of ultrasonic liver images. *IEEE Transactions on Medical Imaging*, 11(2):141–152, June 1992. Cited on page(s) 41

[73] A. N. Esgiar, R. N. G. Naguib, B. S. Sharif, M. K. Bennett, and A. Murray. Fractal analysis in the detection of colonic cancer images. *IEEE Transactions on Information Technology in Biomedicine*, 6(1):54–58, March 2002. DOI: 10.1109/4233.992163 Cited on page(s) 41

[74] T. K. Lee, D. I. McLean, and M. S. Atkins. Irregularity index: A new border irregularity measure for cutaneous melanocytic lesions. *Medical Image Analysis*, 7:47–64, 2003. DOI: 10.1016/S1361-8415(02)00090-7 Cited on page(s) 41

[75] K. Kikuchi, S. Kozuma, K. Sakamaki, M. Saito, G. Marumo, T. Yasugi, and Y. Taketani. Fractal tumor growth of ovarian cancer: sonographic evaluation. *Gynecologic Oncology*, 87:295–302, 2002. DOI: 10.1006/gyno.2002.6842 Cited on page(s) 41

[76] S. H. Nam and J. Y. Choi. A method of image enhancement and fractal dimension for detection of microcalcifications in mammogram. In *Proceedings of the 20th Annual International Conference of the IEEE Engineering in Medicine and Biology Society*, pages 1009–1011, Hong Kong, 1998. Cited on page(s) 41

[77] W. Klonowski, R. Stepien, and P. Stepien. Simple fractal method of assessment of histological images for application in medical diagnostics. *Nonlinear Biomedical Physics*, 4(7):http://www.nonlinearbiomedphys.com/content/4/1/7, 2010. DOI: 10.1186/1753-4631-4-7 Cited on page(s) 41

[78] M. Tambasco, B. M. Costello, A. Kouznetsov, A. Yau, and A. M. Magliocco. Quantifying the architectural complexity of microscopic images of histology specimens. *Micron*, 40(4):486–494, 2009. DOI: 10.1016/j.micron.2008.12.004 Cited on page(s) 41

[79] B. Dubuc, C. Roques-Carmes, C. Tricot, and S. W. Zucker. The variation method: a technique to estimate the fractal dimension of surfaces. In *Proceedings of SPIE, Volume 845: Visual Communication and Image Processing II*, volume 845, pages 241–248, 1987. DOI: 10.1117/12.976511 Cited on page(s) 44, 48

[80] R. S. Weinstein and S. Majumdar. Fractal geometry and vertebral compression fractures. *Journal of Bone and Mineral Research*, 9(11):1797–1802, 1994. DOI: 10.1002/jbmr.5650091117 Cited on page(s) 44

[81] R. C. Coelho, R. M. Cesar Junior, and L. F. Costa. Assessing the fractal dimension and the normalized multiscale bending energy for applications in neuromorphometry. In *Proceedings Simpósio Brasileiro de Computação Gráfica e Processamento de Imagens (SIBGRAPI-96)*, pages 353–554, Caxambu, Brazil, November 1996. Cited on page(s) 44, 51

[82] R. Sedivy, Ch. Windischberger, K. Svozil, E. Moser, and G. Breitenecker. Fractal analysis: An objective method for identifying atypical nuclei in dysplastic lesions of the cervix uteri. *Gynecologic Oncology*, 75:78–83, 1999. DOI: 10.1006/gyno.1999.5516 Cited on page(s) 44

[83] Department of Mathematics and Statistics at Boston University, http://math.bu.edu/ DYSYS/chaos-game/node6.html/. *Fractal Dimension*, accessed October, 2004. Cited on page(s) 48

[84] E. Anguiano, M. A. Pancorbo, and M. Aguilar. Fractal characterization by frequency analysis: I. Surfaces. *Journal of Microscopy*, 172:223–232, 1993.
DOI: 10.1111/j.1365-2818.1993.tb03416.x Cited on page(s) 52

[85] M. Aguilar, E. Anguiano, and M. A. Pancorbo. Fractal characterization by frequency analysis: II. A new method. *Journal of Microscopy*, 172:233–238, 1993.
DOI: 10.1111/j.1365-2818.1993.tb03417.x Cited on page(s) 52

[86] G. D. Tourassi, D. M. Delong, and C. E. Floyd Jr. A study on the computerized fractal analysis of architectural distortion in screening mammograms. *Physics in Medicine and Biology*, 51(5):1299–1312, 2006. DOI: 10.1088/0031-9155/51/5/018 Cited on page(s) 52

[87] R. O. Duda, P. E. Hart, and D. G. Stork. *Pattern Classification*. Wiley, New York, NY, 2nd edition, 2001. Cited on page(s) 55, 56, 59

[88] C. E. Metz. Basic principles of ROC analysis. *Seminars in Nuclear Medicine*, VIII(4):283–298, 1978. DOI: 10.1016/S0001-2998(78)80014-2 Cited on page(s) 61

[89] J. C. Bailar III and F. Mosteller, editors. *Medical Uses of Statistics*. NEJM Books, Boston, MA, 2nd edition, 1992. Cited on page(s) 63

[90] R. E. Walpole, R. H. Myers, and S. L. Myers, editors. *Probability and Statistics for Engineers and Scientists, Sixth Edition*. Prentice Hall, Upper Saddle River, NJ, 1998. Cited on page(s) 63

[91] D. R. Chen, R. F. Chang, C. J. Chen, M. F. Ho, S. J. Kuo, S. T. Chen, S. J. Hung, and W. K. Moon. Classification of breast ultrasound images using fractal feature. *Clinical Imaging*, 29(4):235–245, 2005. DOI: 10.1016/j.clinimag.2004.11.024 Cited on page(s) 82

[92] G. N. Lee, T. Hara, and H. Fujita. Classifying masses as benign or malignant based on cooccurrence matrix textures: A comparison study of different gray level quantizations. In *Digital Mammography: Lecture Notes in Computer Science, Volume 4046/2006*, pages 332–339. Springer, 2006. DOI: 10.1007/11783237_45 Cited on page(s) 83

[93] T. Mu, A. K. Nandi, and R. M. Rangayyan. Analysis of breast tumors in mammograms using the pairwise Rayleigh quotient classifier. *Journal of Electronic Imaging*, 16(4):043004, October 2007. DOI: 10.1117/1.2803834 Cited on page(s) 86

[94] T. Mu, A. K. Nandi, and R. M. Rangayyan. Classification of breast masses via nonlinear transformation of features based on a kernel matrix. *Medical and Biological Engineering and Computing*, 45:769–780, 2007. DOI: 10.1007/s11517-007-0211-0 Cited on page(s) 86

[95] T. Mu, A. K. Nandi, and R. M. Rangayyan. Classification of breast masses using selected shape, edge-sharpness, and texture features with linear and kernel-based classifiers. *Journal of Digital Imaging*, 21(2):153–169, June 2008. DOI: 10.1007/s10278-007-9102-z Cited on page(s) 86

AUTHORS' BIOGRAPHIES

THANH MINH CABRAL

Thanh Minh Cabral obtained her B.Sc. and M.Sc. degrees from the Department of Electrical and Computer Engineering, University of Calgary, Calgary, Alberta, Canada, in 2004 and 2010, respectively. Her research interests are in the areas of medical image processing and computer-aided diagnosis. She has held research and development positions in several engineering companies.

RANGARAJ MANDAYAM RANGAYYAN

Rangaraj Mandayam Rangayyan is a Professor with the Department of Electrical and Computer Engineering, and an Adjunct Professor of Surgery and Radiology, at the University of Calgary, Calgary, Alberta, Canada. He received a Bachelor of Engineering degree in Electronics and Communication in 1976 from the University of Mysore at the People's Education Society College of Engineering, Mandya, Karnataka, India, and a Ph.D. in Electrical Engineering from the Indian Institute of Science, Bangalore, Karnataka, India in 1980. His research interests are in the areas of digital signal and image processing, biomedical signal analysis, biomedical image analysis, and computer-aided diagnosis. He has published more than 140 papers in journals and 250 papers in conference proceedings. His research productivity was recognized with the 1997 and 2001 Research Excellence Awards of the Department of Electrical and Computer Engineering, the 1997 Research Award of the Faculty of Engineering, and by appointment as a "University Professor" in 2003, at the University of Calgary. He is the author of two textbooks: *Biomedical Signal Analysis* (IEEE/ Wiley, 2002) and *Biomedical Image Analysis* (CRC, 2005). He has coauthored and coedited several other books, including one on *Color Image Processing with Biomedical Applications* (SPIE, 2011). He was recognized by the IEEE with the award of the Third Millennium Medal in 2000, and was elected as a Fellow of the IEEE in 2001, Fellow of the Engineering Institute of Canada in 2002, Fellow of the American Institute for Medical and Biological Engineering in 2003, Fellow of SPIE: the International Society for Optical Engineering in 2003, Fellow of the Society for Imaging Informatics in Medicine in 2007, Fellow of the Canadian Medical and Biological Engineering Society in 2007, and Fellow of the Canadian Academy of Engineering in 2009. He has been awarded the Killam Resident Fellowship 3 times (1998, 2002, and 2007) in support of his book-writing projects.

Printed in the United States
by Baker & Taylor Publisher Services